WETLAND PLANTS OF
OREGON & WASHINGTON

WETLAND PLANTS OF
OREGON &
WASHINGTON

by B. JENNIFER GUARD

with contributions from JOHN CHRISTY

photos by TRYGVE STEEN

© 1995 by B. Jennifer Guard
Photography introduction © 1995 by Trygve Steen

First printed in 1995   5 4 3 2
Printed in Canada

The Publisher: *Lone Pine Publishing*

1901 Raymond Ave. SW, Suite C	202A, 1110 Seymour Street	206, 10426 – 81 Ave.
Renton, Washington	Vancouver, British Columbia	Edmonton, Alberta
USA 98055	Canada V6B 3N3	Canada T6E 1X5

Canadian Cataloguing in Publication Data
Guard, B. Jennifer, 1952–
    Wetland plants of Oregon and Washington

    Includes bibliographical references and index.
    ISBN 1-55105-060-9

    1. Wetland plants—Oregon—Identification. 2. Wetland
plants—Washington (State)—Identification. 3. Wetland plants—
Oregon—Pictorial works. 4. Wetland plants—Washington
(State)—Pictorial works. I. Steen, Trygve, 1940– II. Title.
QK144.G82 1995   581.9795   C95-910861-0

Senior Editor: *Nancy Foulds*
Editor: *Roland Lines*
Design: *Beata Kurpinski*
Layout and Production: *Beata Kurpinski, Greg Brown, Bruce Timothy Keith*
Separations and Film: *Elite Lithographers, Edmonton, Alberta*
Printing: *Quebecor Jasper Printing, Edmonton, Alberta*
Cover Design: *Beata Kurpinski*
Cover Photos: *Trygve Steen*

Photographs and illustrations in this book are reprinted by the generous permission of their copyright holders. Illustrations from *Vascular Plants of the Pacific Northwest*, by C.L. Hitchcock, A. Cronquist, M. Ownbey and J.W. Thompson are reprinted by permission of the University of Washington Press. © 1955, 1959, 1961, 1964 and 1969 by the University of Washington Press. Illustrations by Shirley Salkeld are from the collection of and are reprinted by permission of the British Columbia Ministry of Forests Research Branch. © 1993 B.C. Ministry of Forests Research Branch.

Photograph and illustration credits
All photographs are by Trygve Steen, except as follows:
Ed Alverson: p. 121 A & B, p. 131 B, p. 133 and p. 135 B; Hugh G. Barton: p. 152 C and p. 158 B; Jim Harris: p. 195 C; Brett Matthews: p. 147 A, p. 185 B and p. 208 B.
All illustrations are from *Vascular Plants of the Pacific Northwest*, except as follows:
Steve Baker, p. 226 (top 3), p. 227 (top r.), p. 230 (top 2) and p. 231 (bottom r.); Linda Kershaw, p.191 D, p. 226 (bottom), p. 227 (bottom 2), p. 228 and p. 231 (upper 2); Roland Lines, p. 227 (top l.) and p. 231 (bottom l.); John Maywood, p. 56; Shirley Salkeld, p. 229 (bottom) and p. 230 (bottom).

Funding for this publication was made possible through the Oregon Division of State Lands, the Environmental Protection Agency and the United States Senate.

The publisher gratefully acknowledges the support of Alberta Community Development, the Department of Canadian Heritage and the Canada/Alberta Agreement on the cultural industries.

To my son, Peter,
and to my friend, Dennis Michael Holloran,
with whom I have walked and studied many wondrous wetlands,
and who is now undoubtedly mapping the soils of the cosmos.

CONTENTS

FOREWORD

The beauty and variety of the Pacific Northwest is a treasure to all of us who live here. The broad sweep of arid plains, abrupt volcanic peaks, verdant agricultural valleys and rugged coast strike our imagination and shape our images of the Northwest. The damp depressions in the landscape offer a subtle beauty that is often overlooked. This book was conceived as a way to open a window to the beauty of soggy places in Oregon and Washington.

My own experience with the wetlands of the Pacific Northwest started with the tule marshes of the Klamath Basin and wading in the mud of the salt marshes of Yaquina Bay as a child. I have come to know the riparian forests and backwater swamps of alluvial rivers and the wet prairies of the low valleys of the Willamette and Umpqua rivers. I have come to love these landscapes and to appreciate their beauty and value to the wildlife and people of the region.

The photographs in this book tell of the beauty available to us all if we look close; the descriptions help you understand more of what you see. Some of us who have wet feet will recognize good friends in the pictures of this book. I hope the photographs and descriptions will help to open others to the subtle beauty available everywhere in this region. As public-policy battles rage around wetlands, it is important to see and know what the battlefield is.

It is my hope that this book is useful to botanists, environmentalists and anyone interested in the wetlands of the Pacific Northwest. I know it will be useful for wetland professionals and it will become an indispensible part of their toolkit. The focus of this book on the wetland plants of western Oregon and Washington is unique, as is its emphasis on the visual characteristics of those plants. Use it with pleasure and an open mind.

Kenneth F. Bierly
Wetlands Program Manager
Oregon Division of State Lands

PREFACE

This book had its beginnings in a biochemistry research project that examined the mechanisms of heavy metal uptake in aquatic vegetation. Plants contain the menu of elements present in soil and water, whether natural or introduced by human sources, and plants and plant communities provide information relating to ecological and climatic conditions, and they are also indicators of soil types and hydrological states. While identifying plants for the project, it became clear that there was a need for a field guide to common wetland plants.

The organization and contents of this book are intended to help educate people about the important functions and values of wetlands—their importance not only to our water quality but to the general quality of life. The species included in this book are organized by plant communities to impart an understanding of how the unique assemblages of wetland plants, animals, soils and hydrology work together as living systems.

This illustrated wetland guide has been designed with many users in mind—scientists (botanists, limnologists, wildlife biologists and ecologists), private and public agencies that study and manage wetlands, amateur botanists and naturalists, wildflower enthusiasts and all who appreciate, enjoy, study, protect and manage wetlands. The information is presented in a user-friendly, informal and fairly non-technical format. Technical terms are avoided except when required for accuracy, and they are defined in the illustrated glossary. Regardless of your botanical background, you will be able to quickly and accurately identify wetland plants by comparing plants in the field to those in this guide.

B. Jennifer Guard

ACKNOWLEDGMENTS

This book had many diligent contributors, and I wish to thank the following people and establishments for their involvement and support:

Ed Alverson, John Dowd, Judy Elli, Auna Christine Faubion, Jim Harris, Salix & Associates, Dr. Herb Huddleston, Bruce Newhouse, David Liberty, Brett Matthews, Portland State Press, Malcolm H. Scott, The Oregon Natural Heritage Program, Erin Koss, Dr. Stan Swank, Dr. David Wagoner, Dr. La Rea J. Dennis, Emerald Data Base, University of Oregon Map Library staff, Jean Siddall, Neil Björklund, Hugh Barton, Wilber Ternyik, and the University of Oregon Press.

My heartfelt gratitude and appreciation go to the following people:

Senator Mark O. Hatfield, who immediately understood the worthiness of our project and set in motion the support and final funding needed to complete the work, his support and farsightedness are greatly appreciated

Professor Clyde Calvin, PhD, my first botany professor, who anointed my eyes to the mysteries and beauty of plant biology, for his excitement and enthusiasm, which captivated my attention, and for his encouragement and his belief that I could become a botanist

Professor Ed Lippert, PhD, who taught me the many wonders of science, botany and natural history

Professor John Reuter, PhD, who helped me see the need for this book and incubated my ideas for it, for his early support and direction

Nancy Diaz, for providing my earliest opportunity to study wetlands and for the model she provided me of a professional woman ecologist

Gordon Whitehead, for imparting to me the esoteric, beautiful world of sedges, rushes and grasses, and for being an extraordinary person

Chris Orsinger, whose organizational hand nudged the project along in its early phases

Steve Baker, who designed the icon and glossary drawings and provided a touch of class

Ester Lev, for her contributions to the natural history sections of the plant descriptions, not to mention my admiration for all that she has done in our region to promote, educate and help us plan for a future with green spaces, wildlife corridors, healthy rivers, natural landscapes, and to sow within us a growing appreciation of our precious natural resources

Linda Kunze, for her final review of the manuscript

Loverna Wilson, for her final review of the manuscript and affectionate personal support

Peter Zika, for his early reviews of the manuscript and his crazy ways

Professor Trygve Steen, PhD, for the dedicated hours we spent together in the field to capture the exquisite photos of these plants

Ken Bierly, for funding, materials and letting me 'run with the ball,' for his inspiration and constant support, and for being a truly great Wetlands Program Manager to the State of Oregon

John Christy, for his exceptional knowledge of the wetlands of the Pacific Northwest and his generosity with that knowledge, and for his tireless edits and long hours on this project

Scott Sundberg, PhD, for his most generous offerings of time and botanical expertise, and for the resulting suggestions, corrections and insights to the manuscript

Carolyn Rexius, my good friend

Ann Kincheloe, for being the greatest sister anyone could have

Bob Guard, for all his immense love and support at home.

B. Jennifer Guard

INTRODUCTION

Wetland Ecosystems

What is a wetland?

In 1972, by Congressional action, the Army Corps of Engineers was given the authority to regulate the discharge of dredged or fill material into the 'navigable waters of the United States.' Although not obvious in its official wording, this edict included wetlands. The Corps developed a description for wetlands which is still used today—wetlands are 'those areas that are inundated or saturated by surface or groundwater at a frequency and duration sufficient to support and under normal circumstances do support a prevalence of vegetation typically adapted for life in saturated soil conditions.'

In general, wetlands are areas that are transitional between terrestrial and aquatic ecosystems, where the water table is usually at or near the surface or the land is covered by shallow water. The main feature that most wetlands have in common is soil or substrate that is at least periodically saturated with or covered by water. Water is the dominant factor determining the nature of soil development and the types of plant and animal communities living in the soil and on its surface.

There are many general terms for wetlands (marsh, bog, fen, carr, swamp) and specific terms for wetland types (salt marsh, freshwater marsh, tidal flat, peat bog, cranberry bog, scrub shrub swamp, forested wetland), and many different water regimes are assigned to these systems (temporarily flooded, seasonally flooded, semipermanently flooded, intermittently exposed and permanently flooded).

The following three attributes must be found together to determine the existence of a jurisdictional (regulated) wetland:

1 at least periodically, the land supports predominantly hydrophytic (water-loving) vegetation

2 the substrate is predominantly undrained hydric (damp) soil, as indicated by the presence of features such as low or gleyed soils, soft iron masses, oxidized root channels, manganese dioxide nodules, etc.

3 under normal circumstances there is inundation or saturation for two weeks or more during the growing season.

Ownership of land or water alone does not secure ownership of the land or water associated with a wetland. Wetland management attempts to protect wetlands so that humans can continue to enjoy the ecosystem services and resources they provide. The United States Congress, recognizing that over 50% of the nation's original wetlands have been lost to clearing, filling, draining and flood control,* has passed legislation to conserve this valuable resource. Many regulatory bodies are involved in interpreting and enforcing the statutes. Wetland conservation legislation has slowed but not halted the degradation of wetlands.

Wetland functions

For many years, wetlands were regarded as 'wastelands'—without value unless they were used as refuse dumps, drained for agriculture or filled for building construction—but recent research has led to a greater appreciation of the many important ecological functions that wetlands perform.

Functioning wetlands are important natural resources that provide many ecosystem services. The following are some of the most important wetland functions and values:

---

* When the United States was founded there were 86 million hectares (215 million acres) of wetlands; fewer than 40 million hectares (99 million acres) remain.

Water filtration: Ultimately, wetlands are the guardians of our water quality, which they protect by processes of chemical detoxification. The three main processes by which wetlands purify water are as follows:

1 dense wetland vegetation traps and consumes pollutants or buries them in the sediments

2 biochemical processes within a wetland convert harmful chemicals into less harmful ones

3 some pollutants may be taken up by plants and recycled or transported out of the wetland.

Flood control: Wetlands act as natural storm-water run-off and flood-prevention systems by absorbing and then gradually releasing large quantities of water that would otherwise flow quickly downstream. Moreover, the water stored in wetlands during wet periods may be released slowly to adjacent streams during drier periods. This slow release maintains stream flows throughout the summer, which is important for the survival of animals, plants and other organisms that live in or near a stream.

Groundwater recharge: Groundwater is often used as public or private water supply. Much of the water stored by wetlands moves into the subsurface soil. This movement of surface water into the groundwater system is known as groundwater recharge.

Shoreline stabilization: Wetland vegetation binds soil along shorelines and helps reduce erosion caused by wave action.

Fisheries nurseries: Wetlands that surround open water provide habitat for fish spawning nurseries, which improves the health of a stream, fortifies the food chain and provides economically important commercial and recreational fish stocks.

Recreational, aesthetic and cultural values: Wetlands add aesthetic and spiritual richness to our culture through the active and passive recreational values they provide. Birdwatching and nature photography, for example, are popular activities that combine aesthetic, spiritual and recreational components.

Habitat stress

Wetland habitats are both dynamic and demanding. In streams, run-off and floating debris from winter storms routinely rework sediments, destroying old habitat and creating new habitat for aquatic plants. Human activities increasingly determine which species of plants and animals will survive and which ones will disappear.

Depending on local land use and the presence of flood-control dams within a watershed, water levels in wetlands can fluctuate wildly. Bodies of standing water often become turbid after storms. Marshes or ponds not fed by streams typically draw down by as much as fifteen feet in summer, often exposing mud flats or drying up completely. Seasonally, some species must therefore be able to survive both inundation and long periods of exposure. Most of these aquatic systems are also subject to high water temperatures, high nutrient levels and low oxygen in summer.

Life can be rigorous for wetland plants, and many have adapted to these rigors by being opportunists. Most are able to reproduce vegetatively by fragmentation, and when they are submerged, many self-pollinate and set seed without ever opening their flowers. Dispersal is accomplished by seeds or plant fragments moving in water and sediment, as well as seeds eaten by or adhered to birds and mammals. Many wetland species set large quantities of seed that can remain viable for many years.

As in upland habitats, aquatic systems also have their weeds, some of them widespread and serious pests. Aquatic weeds have penetrated nearly all watersheds in the Pacific Northwest, and they can be found at all but the highest elevations. They degrade native plant communities by displacing native species and the animal communities that depend on them.

B. Jennifer Guard
John Christy

Geographical and Species Coverage

Geography was the main guide for the selection of species for this book. The geographic range spans two of four floristic provinces and at least seven geologic provinces of the Pacific Northwest. These areas extend from the Puget Trough province in western Washington through the Willamette Valley province in Oregon, and include representatives from the Umpqua and Rogue river valleys. These areas are broad structural depressions created by the floodplains of the major rivers (the Columbia and the Willamette) and are oriented north and south between the Coast Ranges on the west and the Cascade Range on the east. The Puget Trough extends the entire length of Washington from the Canadian border to Oregon, where the Willamette Valley is its physiographic and geologic continuation. The Willamette Valley, which receives a major focus in this book, is approximately 200 kilometers long and extends from the Columbia River to a transitional point between Eugene and Roseburg, in Oregon, from where the Umpqua and Rogue rivers extend nearly to California.

There is significant overlap of species from province to province, and many of the species described in this book can be found in the Coast Ranges province, the Western Cascades province, the Olympic Peninsula province and in wetlands throughout western Washington and Oregon and into northern California. The regional character of this book reflects the unique assemblages of plants that occur in this widely defined geographic area.

Botanists and wetland researchers throughout the region were consulted about the selection of species for this book. A goal in selecting species was to present a group of species ranging from the truly aquatic to those from a variety of wetland types, as well as species that grow in transitional habitats between wetlands and uplands. While truly aquatic plants are often obvious wetland species, many are included here for the purpose of research and from a general desire to identify these plants. Similarly, when establishing wetland/upland boundaries for the purpose of identifying, studying and protecting wetlands, many terrestrial plants that occur in the transitional areas on or near this boundary become important indicator species.

Unfortunately, it was not feasible to include all the wetland, aquatic and transitional species in the region. This book includes only a fraction of the wetland plant communities that have been described in Oregon and Washington—there are 40 community types described from the Willamette Valley alone. While 155 plant species are described in detail, more than 330 species are discussed or represented in various ways. The 'Similar Species' category of each main entry describes related or 'look-alike' species, often with accompanying photographs.

Generally, this book describes the plants that are dominant or common in wetland communities, although some rare species, such as howellia (*Howellia aquatilis*), have been included. Both native and introduced species are included. To exclude the non-native or exotic species would have left large gaps in identifying plants that occur in wetlands, indicate upland/wetland boundaries and often dominate disturbed wetlands.

The main emphasis of this guide is on flowering plants. However, some ferns, quillworts, horsetails, mosses, liverworts and algae that can be confused with some of the smaller aquatic flowering plants are also described.

Organization of the Book

Our classification of regional wetland types is based on a combination of elements from two main sources—the Oregon Natural Heritage Program's wetland classification (Christy 1993) and the Cowardin classification (Cowardin et al. 1979).

The general habitat types around which the chapters are organized were based on easily observed elements in the landscape, and on groupings of plants and plant communities that occur in or are adapted to distinct water regimes and have similar growth forms. The chapters are ordered for a progression through the book from the most aquatic to the most terrestrial wetland systems.

Physiographic Units

- Coast Ranges
- Siskiyou Mountains
- Western Cascades and Crest
- Eastern Cascades
- Basin and Range
- Owyhee Uplands
- Western Oregon Interior Valley
- High Lava Plains
- Ochoco, Blue and Wallowa Mtns
- Columbia Basin
- Olympic Peninsula
- Puget Trough
- Insular Mountains
- Georgia Depression
- Okanogan Highlands

Each chapter presents a few of the most common plant communities occurring in a particular habitat. The assortment of wetland plant communities appropriate to the geographic context of this field guide were derived from Christy (1993), and they form the basis of wetland plant communities that are characterized here.

Submerged and Floating Communities describes the most truly aquatic plants, which are adapted to life totally in or on top of water.

Marshy Shore Communities moves us out of the water to the muddy shores of lakes, ponds, sloughs, lagoons and streams, where one finds mostly amphibious species that can live both in water and on muddy substrates.

Prairie Wetland Communities is the largest chapter, since the prairie wetlands of the Northwest are endowed with a rich diversity of plant species. It has a special focus on the unique and vanishing Willamette Valley wetland community.

Shrub Swamp Communities takes us to the more terrestrial habitats of willow thickets and more thorny destinations. It focuses on woody wetland plants are less than six meters tall at maturity, and on the plants that are known to favor growing in association with these wetland shrubs.

Wooded Wetland Communities are the most obviously terrestrial, and perhaps the least obvious wetlands to the unacquainted eye. Cottonwood/dogwood stands, ash swales and cedar swamps are highlighted in this chapter, which focuses on woody wetland plants that grow to six meters tall or taller, and on smaller plants that most often associate with them.

Cowardin et al. (1979) was also consulted, primarily because it is the prevailing wetland classification system used in the western United States. The Cowardin system sorts wetlands by a classification hierarchy, with the broadest level divided into the following five systems: marine, estuarine, riverine, lacustrine and palustrine. Cowardin described the level of systems as 'a complex of wetlands and deepwater habitats that share the influence of similar hydrologic, geomorphologic, chemical, or biological factors.'

The five habitat types used to organize this book, and the plants and plant communities grouped within them, are derived from Cowardin's palustrine system, which he describes as follows:

'The Palustrine System was developed to group the vegetated wetlands traditionally called by such names as marsh, swamp, bog, fen, and prairie, which are found throughout the United States. It also includes the small, shallow, permanent water bodies often called ponds. Palustrine wetlands may be situated shoreward of lakes, river channels or estuaries; on river floodplains; in isolated catchments, or on slopes. They may also occur as islands in lakes or rivers.'

The limits of the palustrine system are bounded by upland or by any of the other four systems. Within the palustrine system (as in the four other systems) Cowardin's hierarchy orders the wetlands into classes based on the dominance of vegetation or the substrate, such as rocks, cobble, gravel, sand and mud. Five of eight classes were borrowed from Cowardin's palustrine system and appropriately modified to distinguish and accommodate the generalized habitats used to organize this book.

Cowardin's classification	Equivalent classification in this book
Aquatic Bed	Submerged and Floating
Unconsolidated Shore	Marshy Shore
Emergent Wetland	Wetland Prairie
Scrub/Shrub Wetland	Shrub Swamp
Forested Wetland	Wooded Wetland

The schematic cross-section below shows a typical demarcation of six of the eight classes of Cowardin's palustrine system—including two modifiers of the emergent wetland class, persistent and non-persistent—and their correspondence to the chapters of this book.

Many researchers and professionals in the field were consulted to identify a framework that would both appropriately represent plant communities and provide a friendly, workable structure for presenting the materials within the book.

In the field, the boundaries between these different habitats are not clearly drawn, and many wetlands include several habitat types. For example, a tufted hairgrass prairie usually contains some brush prairie and sometimes a marsh, and a slough along a major river may have a submerged and floating component and a marshy shore, and it may be surrounded by a wooded wetland.

Many wetland plants are perfectly at home in more than one plant community, and especially in more than one of the general habitat types (see appendix 1). For this reason, the habitat icons and frequency bars were developed to help users track the frequency of overlap between habitats.

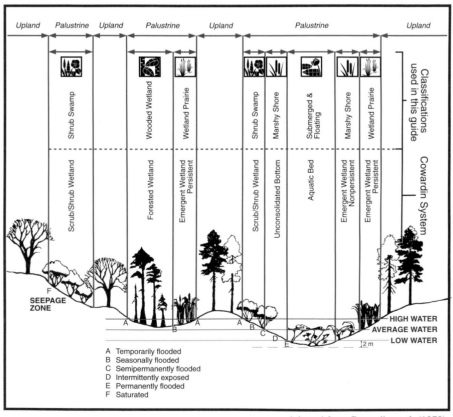

Adapted from Cowardin et al. (1979).

Keys to Identification

Five identification keys help sort out the large and confusing groups of plants. The keys to the rushes (*Juncus* spp.) and the sedges (*Carex* spp.) and the conspectus to the willows (*Salix* spp.) treat large genera that are richly represented in wetland habitats. The key to the grasses and the key to the pondweeds (*Potamogeton* spp.) and other aquatic species treat groups of species from several genera linked by similar habitats and growth forms.

The keys contain as little technical jargon as possible, but the use of some conventional diagnostic terminology well known to botanists was necessary for accurate identifications. Most unfamiliar terms will be found in the illustrated glossary.

Some species treated by the keys may not be found in this guide or are only casually mentioned. The development of these keys was another chance to broaden the scope of this book where space confinements prevented the addition of a species to the main descriptive section of the book. This project continually looked for ways of treating as many plants as possible to reduce the chance that the plant in the hand of the user would not be identifiable.

The Plant Descriptions

Species names

To make this book accessible to a wide audience, the plants are primarily referred to by their common names. The common names have generally been taken from *Vascular Plants of the Pacific Northwest* (Hitchcock et al. 1955–69), but many species have more than one common name, and all other common names that were encountered during the research of this book are also given.

Vascular Plants of the Pacific Northwest was also used for most of the scientific nomenclature. However, the recent publication of Hickman's *Jepson Manual: Higher Plants of California* (1993) made it possible to update some of the outdated nomenclature of the Pacific Northwest. In addition, the Oregon Flora Project at Oregon State University contributed generously to the improved nomenclature of many species treated and mentioned here.

Synonymous and misapplied scientific names are clearly listed, since many synonyms are still widely used, and an earnest attempt to connect the correct names to the synonyms has been made.

The family name is listed for each species to provide additional taxonomic information that links species at a higher taxonomic level. Some family names have undergone recent changes, and some genera have shifted to different families.

Indicator status

This is a reference for people who use this guide with the federal methodology for delineating jurisdictional wetlands. The indicator status assignments are based entirely on the *National List of Plant Species That Occur in Wetlands* (Reed 1988), and they are defined as follows:

Obligate Wetland (OBL): A species that almost always occurs under natural conditions in wetlands (estimated probability greater than 99%).

Facultative Wetland (FACW): A species that usually occurs in wetlands (estimated probability 67%–99%), but is occasionally found in non-wetlands.

Facultative (FAC): A species that is equally likely to occur in wetlands or non-wetlands (estimated probability 34%–66%).

Facultative Upland (FACU): A species that usually occurs in non-wetlands (estimated probability 67%–99%), but is occasionally found in wetlands (estimated probability 1%–33%).

Obligate Upland (UPL): A species that occurs in wetlands in another region (estimated probability greater than 99%), but almost always occurs under natural conditions in non-wetlands in the region specified (region 9 for this book). If a species does not occur in wetlands in any region, it is not on the national list (NOL).

Key to a Sample Entry

habit icon for this chapter

common name   **alternate common name**

WATSON'S WILLOW-HERB • HAIRY WILLOW-HERB

current scientific name *Epilobium ciliatum ssp. watsonii (E. watsonii)*

secondary habitat icons (with frequency bars)

family name ▸ Onagraceae (Evening-primrose Family)

INDICATOR STATUS: FACW-

former scientific name

description of main species

GROWTH HABIT: Perennial native forb; typically **30–100 cm tall**; spreads by short underground stems (rhizomes); main stems branch in upper half, especially in flower clusters.

LEAVES: Opposite, densely arranged on stem, strongly veined, **glossy green, lance- to egg-lance-shaped, 1–15 cm long**, stalkless (**sessile**) or nearly so (**subsessile**), appear to attach directly to stems; **edges minutely toothed**.

FLOWERS: In loose, **compound clusters (inflorescences); pink to rose-purple; four petals, 5–14 mm long**, light pink, **notched at edges**; fruit forms below flower (inferior ovary).

HABITAT: Shallow, fresh water, including marshes, shrub swamps and wet meadows, especially in disturbed habitats.

NATURAL HISTORY: Watson's willow-herb is often found in the understory of cattail (*Typha latifolia*), sedges (*Carex* spp.) and rushes (*Juncus* spp.). It is native to western North America, but it has spread over much of the United States and has invaded Europe.

description of similar species

SIMILAR SPECIES: Smooth willow-herb (*E. glaberrimum* ssp. *glaberrimum*, FACW) has small (1–7 cm long), narrow, opposite leaves that clasp the stem and appear to attach directly to it. Common willow-herb (*E. ciliatum* ssp. *glandulosum*, also known as *E. glandulosum*, FACW-) has fleshy, rosebud-like offshoots (turions) on its roots or rhizomes at maturity. You have to uproot a plant to see this feature. Dense spike-primrose (*E. densiflorum*, p. 118) has opposite, hairy leaves (except those near the stem base), and its stems peel towards the base. The leaves of purple loosestrife (*Lythrum salicaria*, p. 75) attach directly to the stem and have grayish hairs on their upper surfaces. They can appear to be in whorls around the stem, but actually they are opposite. The flowers of purple loosestrife are densely crowded in the upper leaf axils, in an interrupted pattern. Hyssop loosestrife (*Lythrum hyssopifolium*, p. 75) is a pale waxy-bluish green annual with both opposite and alternate leaves (in threes). Also, both species of loosestrife are much larger than Watson's willow-herb.

other interesting notes

NOTES: The willow-herbs belong to the same genus as fireweed (*E. angustifolium*).

photos of the main species (sometimes a similar species is also shown in a photo or illustration)

A, B: *Watson's willow-herb* (Epilobium ciliatum *ssp.* watsonii).

photo/illustration captions

SHRUB SWAMP COMMUNITIES 175

Not Found on List (NOL): A species for which nothing has been reported.

No Indicator (NI): A species for which there is insufficient data to provide an indicator determination.

Non-occurrence in Region (NO): A species that is recorded on the national list, but does not occur in the region specified (region 9 for this book).

No Agreement Reached (NA): A species for which no agreement or consensus regarding its indicator status has been reached by the regional panel.

The addition of **+** or **-** after an indicator status shows a tendency towards the higher (+) or lower (-) end of the given range of probabilities.

A full listing of the indicator status assignments for the species in this book is given in appendix 2.

Habitat icons

Each of the chapters of the book is represented by an icon that serves to identify the general habitat type of the chapter.

 Submerged and Floating

 Marshy Shore

 Wetland Prairie

 Shrub Swamp

 Wooded Wetland

These icons are also used to identify other habitats in which a plant species may be found. The frequency bars below these secondary habitat icons reflect the relative frequency (typically, occasionally or infrequently) with which each plant is found in a habitat type.

Two additional icons identify species that are noxious pests or that occur in sites disturbed by human activity.

 Noxious Pest

 Disturbed Site

Descriptive text

Growth Habit: The first element of this section is included as a reference for people who use this book in association with the federal methodology for delineating jurisdictional wetlands. It is based on the *National List of Plant Species That Occur in Wetlands* (Reed 1988), except where obvious errors were made, such as listing an annual plant as a perennial. As many of these errors were corrected as possible. The rest of this section describes the size, shape and texture and other general characteristics of a plant, including the features of the stems and roots. It is important to distinguish between stems and stalks—true stems can be limited structures on certain plants, and leaf attachments and other stem-like appendages are referred to by their appropriate names in this guide.

Leaves describes the important features of a plant's leaves, which can be particularly useful in identification. Many aquatic plants have dimorphic leaves, which means there are two types of leaf present on the same plant. The leaf forms are usually related to whether the leaf is growing submerged in water or in the air.

Flowers are usually the best structure to have for traditional identification of plants, and this category describes the more obvious and important features of the flower that can aid in identification.

Fruits describes the important features of a plant's fruits and seeds.

Habitat refers, in general, to certain situations or conditions in which a plant lives. Knowing the general environment in which you can expect to find a plant aids in its identification.

Natural History provides a variety of information about a plant's flowering time, its range and abundance, the other plant species with which it is commonly associated, and the various ways it is used by wildlife. Much of the information included here is relevant to wildlife and restoration uses of wetland plants.

Similar Species describes species that can be easily mistaken or confused with the main species discussed. Related species that are not as easily confused, but share some similarities, are also included here. *Flora of the Pacific Northwest* (Hitchcock & Cronquist 1973) is a useful reference for more detailed information about these 'secondary' species.

Notes contains a collage of interesting and useful information, gathered from many sources, including medicinal uses and other ethnobotanical facts, toxicity and edibility, and the origins of some plant names.

Although this guide records medicinal and food uses of plants, it is not a 'how-to' reference for using traditional plant-derived medicines and foods, nor does it advocate the use of such medicines and foods. Experimentation by the reader is also not recommended. It is cautioned that many plants in our region, including some with traditional medicinal uses, are poisonous or harmful.

B. Jennifer Guard

Photography

The priority when photographing plants for this book was to present features that would be most diagnostic for an accurate identification. Specimens were carefully selected to show as many distinguishing characteristics as possible within the area of a single image. Many of the aquatic plants were collected and photographed in a shallow aquarium so they could be seen clearly.

The photographs were taken with a 35 mm single-lens-reflex camera equipped with a 105 mm macro lens and one or more electronic flash units. Careful control of the plane of focus, depth of field, lighting characteristics and exposure enabled each subject's diagnostic features to be effectively captured in the photographs.

The use of electronic flash as the primary light source for this photography provided a number of major advantages—it was easier to attain appropriate depth of field; sharpness was more consistent, especially under breezy conditions; lighting could be controlled so the important structural features could be more effectively shown; and the limited ability of flash to illuminate the background could be used to help isolate the subject.

A 105 mm macro lens was used for most photographs because it provided great image sharpness, an ability to focus to a 1x magnification when needed, a more natural perspective than shorter focal lengths, and sufficient distance between the lens and the subject so that flash could be more effectively used for illumination. Wide-angle lenses were used for habitat shots and occasionally for whole plant shots, while telephoto lenses were used to reach out and capture images of specimens in difficult locations.

Lower speed slide films were chosen for this work because of their fine grain, ability to render detail, image stability and quality of color reproduction. Kodachrome 64 was the most often used film. Fujichrome 50 was used for purple- and blue-flowered specimens because of its ability to render these flower colors more dependably.

Trygve Steen

KEYS TO IDENTIFICATION

Key to the Pondweeds (Potamogeton *spp.*) *and Other Aquatic Species*

1a. Only one leaf type 2

  2a. Leaves 3–30 mm wide 3

    3a. Leaves bright, glossy green with crisped (rippled) edges .... *Potamogeton crispus*

    3b. Leaves with wide, wavy edges, mostly 2–3 mm wide; stems whitish .... *Potamogeton praelongus*

  2b. Leaves narrowly linear, less than 2 mm wide 4

    4a. Leaf bases broad and clasping stem; leaf edges with sharp teeth; flowers solitary in leaf axils 5

      5a. Leaves long, narrow, and tapering to tip .... *Najas flexilis*

      5b. Leaves short and blunt .... *Najas guadalupensis*

    4b. Leaves and flowers not as above 6

      6a. Leaves alternate or opposite 7

        7a. Leaves finely divided 2–3 times (from near base) into branched, finger-like segments, some segments bearing specialized bladders that trap small invertebrates .... *Utricularia macrorhiza*

        7b. Leaves simple 8

          8a. Leaves all opposite; flowers in groups of 2–5 in leaf axils 9

            9a. Leaves hair-like, 2–10 cm long .... *Zannichellia palustris*

            9b. Leaves mostly linear (upper leaves may be spatulate in some species), less than 2 cm long .... *Callitriche* spp.

          8b. Leaves mostly alternate (a few may be opposite); flowers in spike-like clusters at stem tips 10

            10a. Submersed leaves with stipules fused to leaf base and forming a sheath around stem; leaf blades attached to end of stipules, less than 1 mm wide, usually 1-nerved .... *Potamogeton pectinatus*

            10b. Submersed leaves with stipules free from leaf blade; leaf blades attached to stem node, 1–2 mm wide and usually 3-nerved .... *Potamogeton foliosus*

      6b. Some leaves or lobes in whorls of 3 or more 11

        11a. Algae with branching lobes in whorls 12

          12a. Lobes all the same length, gray-green, gritty with lime deposits, having a disagreeable odor .. *Chara* spp.

          12b. Lobes sometimes forked 2–3 times, delicate, bright green, not gritty, no disagreeable odor .. *Nitella* spp.

        11b. Vascular plants with leaves in whorls (lower leaves sometimes opposite) 13

          13a. Leaves all simple and undivided 14

            14a. Leaves in whorls of 3 (lower leaves sometimes opposite), 0.6–1.5 cm long ..... *Elodea canadensis*

            14b. Leaves in whorls of 4–6, mostly 2.5–3 cm long .... *Egeria densa*

          13b. All or some leaves finely pinnately divided and often feather-like 15

            15a. Leaves all feather-like, finely pinnately divided into 25–36 thread-like segments; an introduced species often escaped from aquaria .... *Myriophyllum aquaticum*

            15b. Submerged leaves feather-like, pinnately divided into 13–23 thread-like segments, but gradually transitional upwards to simple linear-oblong, toothed bracts at stem tip ..... *Myriophyllum hippuroides*

1b. Leaves of two types, with broad, floating leaves and narrower, more elongate submerged leaves 16

  16a. Leaves lobed or dissected, wider than long; floating leaves 3-lobed, with toothed or lobed segments; submerged leaves 3–7-times divided into hair-like segments .... *Ranunculus aquatilis*

  16b. Leaves simple, longer than wide 17

    17a. Submerged leaf blades mostly less than 2 mm wide and over 10 cm long; floating leaves heart-shaped at base .... *Potamogeton natans*

    17b. Submerged leaf blades either more than 2 mm wide or less than 10 cm long; floating leaves somewhat tapered at base, not heart-shaped 18

      18a. Submerged leaves 2–5 cm wide, folded lengthwise and curved like a hawk's beak; stipules mostly over 5 cm long .... *Potamogeton amplifolius*

      18b. Submerged leaves mostly less than 2 cm wide and neither folded nor curved; stipules mostly less than 5 cm long 19

        19a. Some stipules at least 3 cm long; submerged leaves often stalked .... *Potamogeton nodosus*

        19b. Stipules less than 3 cm long; submerged leaves stalkless 20

          20a. Plants large, 0.5–1.5 m long; submerged leaves 10–20 cm long, with a broad stripe at least 1 mm wide down the center .... *Potamogeton epihydrus*

          20b. Plants smaller, 0.3–0.6 m long; submerged leaves less than 10 cm long, without a broad central stripe .... *Potamogeton gramineus*

Key to the Grasses

(Separate keys to some of the larger genera follow this main key.)

1a. Ligules lacking .... *Echinochloa crus-galli*

1b. Ligules present 2

  2a. Spikelets break off below glumes, either at base of spikelet or along rachis or branches thereof, shed as a unit . 3

    3a. Ligules mainly a fringe of hairs; spikelets in open panicles; (*Panicum*) 4

      4a. Plants annual; ligules 0.5–1.5 mm long; inflorescence 15–40 cm long .... *Panicum capillare*

      4b. Plants perennial; inflorescence 4–9 cm long .... *Panicum acuminatum*

    3b. Ligules membranous 5

      5a. Glumes lacking; plants perennial, rhizomatous .... *Leersia oryzoides*

      5b. Glumes present 6

        6a. Spikelets sessile or subsessile 7

          7a. Spikelets in spike-like racemes, confined to one side of the rachis, 1–2-flowered
(lower flower sometimes sterile) .... *Beckmannia syzigachne*

          7b. Spikelets in a single terminal spike, borne alternately on opposite sides of the rachis 8

            8a. Spikelets dissimilar, usually 3 per node; central spikelet sessile and fertile;
plants annual or non-rhizomatous perennial .... *Hordeum*

            8b. Spikelets mostly one per node, 4–12-flowered, flattened; plants perennial,
from short rhizomes; primarily an upland species .... *Pseudoroegneria spicata*

        6b. Spikelets stalked in open to congested panicles (not confined to one side of rachis) 9

          9a. Spikelets 1-flowered; panicles condensed and spike-like .... *Alopecurus*

          9b. Spikelets 2–3-flowered, with one or more flowers sterile or staminate 10

            10a. Spikelets 3-flowered; leaves hairless; (*Phalaris*) 11

              11a. Inflorescence sometimes interrupted in lower third; generally egg-shaped;
rhizomes short; upper lemmas densely hairy .... *Phalaris aquatica*

              11b. Inflorescence more or less cylindrical, generally interrupted at base;
rhizomes well-developed, evident; upper lemmas sparsely hairy or hairless .... *Phalaris arundinacea*

            10b. Spikelets 2-flowered; leaves covered with dense, velvety hairs; (*Holcus*) 12

              12a. Stems tufted; nodes and internodes soft and hairy; lemma awn about 1 mm long,
twisted to recurved .... *Holcus lanatus*

              12b. Stems solitary to few, from rhizomes; internodes hairless; nodes with downward-
pointing hairs; lemma awn 3–4 mm long, bent to straight .... *Holcus mollis*

  2b. Spikelets break off above glumes 13

    13a. Ligules mainly a fringe of hairs; plants perennial, 30–80 cm tall;
spikelets all alike, 1–5 per plant .... *Danthonia californica*

    13b. Ligules mainly membranous 14

      14a. Leaf sheaths closed at base for at least one quarter of their length; leaf tips hooded or prow-like (not flat) 15

        15a. Leaves folded in the bud; leaf tips usually prow-like; spikelets stalked in open
to contracted panicles; lemma nerves converging .... *Poa*

        15b. Leaves rolled in the bud; leaf tips not prow-like (but flat and pointed) 16

          16a. Lemma nerves not converging; lemmas unawned; both glumes 1-nerved .... *Glyceria*

          16b. Lemma nerves converging; lemma awns 3–15 mm long; first glume 1- or 3-nerved;
second glume 3- or 5-nerved .... *Bromus carinatus*

      14b. Leaf sheaths mostly open (closed for less than one quarter of their length);
leaf tips flat (not hooded or prow-like) 17

        17a. Spikelets 1-flowered 18

          18a. Culms usually bulbous-based; panicles cylindrical, spike-like, usually over 4.5 cm long;
mid-nerve of glumes extending into an abrupt point; primarily an upland species ...... *Phleum pratense*

          18b. Culms never bulbous-based; panicles often open; glumes mostly gradually tapering 19

             19a. Callus hairs usually at least three quarters as long as lemma; lemmas delicately
awned; paleas well developed, usually nearly as large as lemma .... *Calamagrostis canadensis*

             19b. Callus minutely bearded or with hairs half as long as lemma; lemmas unawned or awned
from back; palea often reduced, usually less than half as long as lemma .... *Agrostis*

        17b. Spikelets 2- to many-flowered 20

          20a. Plants annual; spikelets strongly compressed, about 3 mm long, up to 7-flowered,
broadly triangular, many spikelets per panicle; glumes spreading .... *Briza minor*

          20b. Plants perennial 21

21a. Spikelets 3-flowered; first glume half as long as second; uppermost flower perfect
and with both a lemma and a palea; lower flowers sterile .... *Anthoxanthum odoratum*
21b. Spikelets 2- to many-flowered; reduced flowers (if any) are above or both above and below perfect ones 22
22a. One or both glumes longer than first lemma; lemmas awned from the back; rachilla prolonged
beyond second flower; plants tufted .... *Deschampsia*
22b. Both glumes shorter than first lemma; lemmas unawned, or with awn as minute extension of mid-nerve 23
    23a. Lemmas unawned, with 5–7 parallel (non-converging), fairly prominent nerves;
    primarily a species of higher-elevation wetlands .... *Puccinellia*
    23b. Lemmas often unawned, with usually obscure, converging nerves 24
        24a. Spikelets 2–12-flowered, stalked in open to contracted panicles; leaf blades rolled in bud;
        leaf tips pointed .... *Festuca*
        24b. Spikelets compressed, usually 3-flowered, subsessile, in 1-sided clusters on long, stiff, ascending panicle
        branches; leaf blades folded in bud; leaf tips hooded; primarily an upland species .... *Dactylis glomerata*

Key to the *Agrostis* species

1a. Plants erect .... *A. capillaris*
1b. Plants often decumbent, prostrate and stoloniferous 2
    2a. Panicle generally compressed or contracted; plants to 120 cm tall .... *A. alba*
    2b. Panicle open or moderately so; plants less than 60 cm tall 3
        3a. Panicle reddish or purplish .... *A. diegoensis*
        3b. Panicle greenish, or otherwise not purple or red 4
            4a. Spikelets overlapping and crowded on same branch .... *A. stolonifera*
            4b. Spikelets not crowded, generally well spaced on same branch .... *A. idahoensis*

Key to the *Alopecurus* species

1a. Plants annual; stems generally less than 45 cm tall; lemma awn straight, less than 4 mm long .... *A. aequalis*
1b. Plants perennial; stems generally more than 45 cm tall 2
    2a. Lemma awn straight, more than 4 mm long .... *A. pratensis*
    2b. Lemma awn bent, less than 4 mm long .... *A. geniculatus*

Key to the *Deschampsia* species

1a. Plants annual; panicles open to narrow; ligules 2–4 mm long; blades generally inrolled,
1–9 cm long, 1–2 mm wide .... *D. danthonioides*
1b. Plants perennial 2
    2a. Panicles open, 20–100 cm long, with spreading branches; ligules 3-8 mm long;
    blades 8–20 cm long, 1–4 mm wide .... *D. cespitosa*
    2b. Panicles narrow and appressed to stem; branches spike-like and ascending .... *D. elongata*

Key to the *Festuca* species

1a. Basal lobes of leaf blade, if present, not clasping stem if sheath closed; inflorescence more or less reddish .. *F. rubra*
1b. Basal lobes or leaf blade prominent, more or less clasping stem; inflorescence green,
becoming straw-colored soon after fruits mature 2
    2a. Basal lobes hairy; plants 80–200 cm tall .... *F. arundinacea*
    2b. Basal lobes not hairy; plants 30–130 cm tall .... *F. pratensis*

Key to the *Glyceria* species

1a. Inflorescence narrow, with appressed branches 2
    2a. Lemmas glabrous between barely scabrous veins .... *G. borealis*
    2b. Lemmas minutely scabrous between distinctly scabrous veins; lemma tip jagged,
    longest between veins .... *G. occidentalis*
1b. Inflorescence branches spreading 3
    3a. Palea tip jagged or widely V-notched .... *G. grandis*
    3b. Palea tip narrowly or minutely notched 4
        4a. Leaf blades 2–8 mm wide, firm; lower glumes 0.4–0.8 mm long .... *G. striata*
        4b. Leaf blades 6–10 mm wide, thin; lower glumes 0.8–1.2 mm long .... *G. elata*
            (*G. elata* intergrades with both *G. grandis* and *G. striata*)

Key to the *Hordeum* species

1a. Glumes awn-like, 2–6 cm long; lemma awns nearly as long as glumes; spike (awns included) nearly as broad as long; plants perennial, densely tufted . . . . *H. jubatum*
1b. Glumes often broadened at base, less than 2 cm long; spike (awns included) much longer than broad 2
2a. Plants perennial; stems up to 90 cm tall; spikes 5–10 cm long . . . . *H. brachyantherum*
2b. Plants annual; stems 10–40 cm tall; spikes 3–6 cm long . . . . *H. depressum*

Key to the *Poa* species

1a. Plants annual . . . . *P. annua*
1b. Plants perennial 2
2a. Culms strongly flattened, 2-edged; lemmas scantily if at all webbed; plants strongly rhizomatous; primarily an upland species . . . . *P. compressa*
2b. Culms slightly if at all flattened, not 2-edged; lemmas more or less strongly webbed at base; panicles open, with mostly spreading branches 3
3a. Plants with true rhizomes, usually not growing in wet places; ligules mostly not over 1.5 mm long; lemmas more than 3 mm long . . . . *P. pratensis*
3b. Plants with stolons but not rhizomes, usually growing in wet places; ligules mostly 3–7 mm long; lemmas 2.5–3 mm long 4
4a. Panicles mostly 10–30 cm long, open and loose, with strongly spreading branches; lemmas obscurely nerved . . . . *P. palustris*
4b. Panicles mostly 8–15 cm long, loose, with usually somewhat ascending branches; lemmas more strongly nerved . . . . *P. trivialis*

Key to the Rushes (Juncus spp.)

1a. Plants annual 2
2a. Plants generally less than 4 cm tall; flowering stems unbranched and leafless . . . . *J. hemiendytus*
2b. Plants generally more than 4 cm tall; many branching flowering stems . . . . *J. bufonius*
1b. Plants perennial 3
3a. Leaves iris-like, flattened and overlapping 4
4a. Inflorescence of less than 10 globular heads (each about 10 mm across) . . . . *J. ensifolius*
4b. Inflorescence of 10–70 heads (each about 5 mm across), spreading . . . . *J. oxymeris*
3b. Leaves grass-like, rounded or vestigial 5
5a. Leaves grass-like, flattened and thin 6
6a. Plants growing from spreading rhizomes . . . . *J. covillei*
6b. Plants strongly tufted 7
7a. 3–4 alternate leaves per stem; knobby, bulb-like rhizomes . . . . *J. marginatus*
7b. All leaves near stem base; without knobby, bulb-like rhizomes . . . . *J. tenuis*
5b. Leaves vestigial or rounded (terate) 8
8a. Leaves with cross walls (septate) 9
9a. Inflorescence of a few globular, compact heads usually more than 8 mm across . . . . *J. bolanderi*
9b. Inflorescence of many smaller heads, spreading 10
10a. Inflorescence on a stalk that is longer than lowest inflorescence bract . . . . *J. articulatus*
10b. Inflorescence branching near base of lowest inflorescence bract 11
11a. Plants strongly rhizomatous; perianth longer than capsule . . . . *J. nevadensis*
11b. Plants tufted, with short rhizomes; perianth shorter than or equalling capsule . . . . *J. acuminatus*
8b. Leaves without cross walls (not septate) 12
12a. Inflorescence terminal; leaves with well-developed, flattened blades . . . . *J. tenuis*
12b. Inflorescence appearing lateral; leaves bladeless 13
13a. Stems bluish green, slender, lax or spreading . . . . *J. patens*
13b. Stems bright or dark green, erect 14
14a. Capsule firm; 6 stamens; anthers longer than filaments . . . . *J. balticus*
14b. Capsule easily ruptured; 3 stamens; anthers shorter than or equalling filaments 15
15a. Perianth 2–3.5 mm long, soft when fresh, not rigid when dry, dark brown . . *J. effusus var. gracilis*
15b. Perianth 2.5–4.2 mm long, firm when fresh, rigid when dry, green to pale brown . . . . *J. effusus var. pacificus*

Key to the Sedges (Carex spp.)

1a. Spikes (at least some) with evident stalks; 2 or 3 stigmas, with achenes accordingly lens-shaped or 3-sided 2
  2a. Styles have same bony texture as achenes and do not wither with time, twisted and contorted within perigynia; 3 stigmas; female spikes erect or stiffly ascending 3
    3a. Perigynia ascending, lance- to lance-egg-shaped; stems more or less clustered on short rhizomes . . *C. vesicaria*
    3b. Perigynia strongly spreading at maturity, elliptical or egg-shaped to nearly round; plants sod-forming, with stems arising singly or a few together from long rhizomes . . . . *C. utriculata*
  2b. Style soon withering and breaking from achene 4
    4b. 2 stigmas; achenes lens-shaped 5
      5b. Perigynia more or less membranous; achenes never constricted at middle . . . . *C. aperta*
      5a. Perigynia thick-walled and firm; achenes distinctly constricted at middle 6
        6a. Female spikes 5–12 cm long, stalkless or nearly so, but spike itself is lax, with a nodding tip; leaf sheaths fibrous-shreddy at base . . . . *C. obnupta*
        6b. Spikes shorter (1.5–5 cm long), at least lower ones generally nodding on slender, elongate stalks; leaf sheaths not fibrous-shreddy . . . . *C. lyngbyei*
    4a. 3 stigmas; achenes 3-sided 7
      7a. Perigynia densely velvety; larger leaves mostly 2–5 mm wide . . . . *C. lanuginosa*
      7b. Perigynia hairless 8
        8a. Bracts subtending spikes all sheathless, or nearly so; leaves very broad (8–20 mm wide); stems arising singly or a few together from long, creeping rhizomes . . . . *C. amplifolia*
        8b. At least lowest bract (subtending lowest spike) has a well-developed sheath (1.5–7 cm long); larger leaves 6–14 mm wide; plants tufted, without rhizomes . . . . *C. hendersonii*
1b. Spikes stalkless; 2 stigmas; lens-shaped achenes 9
  9a. Spikes with male flowers at tips and female flowers at base; plants more or less densely tufted (rhizomes absent or very short) 10
    10a. Leaf sheaths generally have conspicuous horizontal puckering or wrinkles 11
      11a. Perigynia lance-triangular, broadest at or near base and tapering gradually and evenly to tip, 4–5.2 mm long . . . . *C. stipata*
      11b. Perigynia broadly egg-shaped, broadest above base and contracted to a distinct, flattened beak 12
        12a. Scales of female flowers tipped with a 1–5-mm-long bristle; perigynia 2–3.5 mm long . . . *C. vulpinoidea*
        12b. Scales of female flowers simply pointed or with a short bristle less than 1.5 mm long; perigynia 3–4 mm long . . . . *C. densa*
    10b. Leaf sheaths not at all puckered or wrinkled 13
      13a. Leaf sheaths not red-dotted; leaves 1–2.5 mm wide; few spikes (mostly 10 or less) loosely aggregated in irregular or oblong-cylindrical head (1.5–3.5 cm long) . . . . *C. tumulicola*
      13b. Leaf sheaths red-dotted or more or less coppery tinged; leaves wider (larger ones mostly 3–5 mm wide); spikes more numerous, in compound, grass-like inflorescence (3–8 cm long) . . . . *C. cusickii*
  9b. Spikes with female flowers at tips and male flowers at base (some spikes may be entirely female) 14
    14a. Perigynia often with raised edges, but never thin-edged; 2–8 spikes crowded into a head; perigynia 3.5–4.8 mm long, relatively long-beaked . . . . *C. deweyana*
    14b. Perigynia with thin edges, or winged 15
      15a. One or more of the lower bracts are equal to or longer than inflorescence 16
        16a. Lowest bract 5–15 cm long, often erect and appearing like a continuation of the stem; beak of perigynium ill-defined . . . . *C. unilateralis*
        16b. Lowest bract 2–8 cm long, bristle-like, not appearing like a continuation of stem; beak of perigynium more evident . . . . *C. athrostachya*
      15b. Bracts of inflorescence are all short or hair-like 17
        17a. Ligule elongate, 3–8 mm long (but mostly joined to blade); ventral side of leaf sheath firm and green all the way to the collar, except sometimes for short, colorless triangle within 5 mm of collar . . . . *C. feta*
        17b. Ligule short (less than 3 mm long) 18
          18b. Pistillate scales larger, more or less completely conceal perigynia; perigynia strongly flattened . . . . *C. leporina*
          18a. Pistillate scales distinctly shorter and narrower than perigynia 19
            19a. Perigynia nearly filled by plump achene; inflorescence 1–2 cm long . . . . *C. pachystachya*
            19b. Perigynia strongly flattened and much wider than achene; inflorescence 2–4 cm long . . . *C. scoparia*

Conspectus of the Willows (Salix spp.)

	S. scouleriana	S. hookeriana	S. sitchensis
Growth habit	shrub or small tree, 2–12 m tall	shrub or small tree, 3–6 m tall	many-stemmed shrub, 2–6 m tall
Leaf shape	oval to lance-shaped, usually broadest at tip, twice as long as wide; tip sometimes sharply to slightly pointed	broadly elliptical, 4–12 cm long, 1–4 cm wide (up to 4 times as long as wide); tip rounded or slightly pointed	broadly lance-shaped, widest near tip, 4–9 cm long, 1.5–3.5 cm wide
Leaf edges	occasionally has rounded, loosely arranged teeth	loosely toothed, somewhat wavy or curled; teeth rounded or scalloped (crenate)	bearing some scattered, small, toughened glands
Leaf stalk	5–10 mm long	0.5–2 cm long	0.5–1.5 mm long
Young leaf surfaces	fine-hairy above; waxy below, with dense white to rust-colored hairs	glossy green above, with dense hairs; waxy, grayish blue (glaucous) below	wrinkled, dull grayish green above; gray or white below, with dense, long, slender, silky, soft, appressed hairs
Old leaf surfaces	glossy green above; prominent veins with silvery-red hairs below	glossy green above, with dense hairs; neither glossy nor bright green below, with dense, white-woolly hairs	dark green above, with thin, sparse hairs, or hairless; dense silvery hairs, like crushed velvet, below
Leaf stipules and glands	stipules inconspicuous and deciduous	well-developed, collar-like stipules at stalk base	leaf-like stipules at base of most leaves, especially on vigorous new shoots
Female catkins	2.5–6 cm long, appearing before leaves; bracts dark brown to black	4–12 cm long; bracts brown to black	3–8 mm long; bracts light brown to black; stalk short (1 cm long) with leafy bracts along it
Male catkins	2–4 cm long, subsessile, or on short (1.5 cm long) stalk	3–4 cm long, 2–2.5 cm wide, subsessile (stalk less than 1 cm long); bracts small and inconspicuous, or up to 3 cm long	2.5–5 cm long, 1–1.5 cm wide; stalk 1 cm long, with 2.5-cm-long, leafy bracts
Twigs	reddish brown (striped) when young; two-toned greenish brown when old	fine-hairy, green when young, gradually becoming coarser with age; dark gray, hairless and smooth by 1 year	densely velvety when young; dark, dull gray and hairless when old

S. lucida ssp. lasiandra	S. geyeriana	S. sessilifolia	S. fluviatilis
large shrub (3–6 m tall) or small tree (to 10 m tall)	shrub, 4–6 m tall	shrub or small tree, 2–8 m tall	shrub or small tree, 2–6 m tall
lance-shaped to narrowly elliptical, 5–15 cm long, 1–3 cm wide, tapering to pointed tip; vigorous new leaves can be quite large (up to 5 cm by 25 cm)	lance-shaped, 3.2–7.4 cm long, 0.5–1.2 cm wide; tip tapered; base wedge-shaped	narrowly lance-shaped, 5–15 cm long, 0.4–1.5 cm wide	narrowly lance-shaped, 5–15 cm long, 0.4–1.5 cm wide
bearing fine, close serrations	shallowly toothed	bearing many scattered teeth	bearing many scattered teeth
3–15 mm long	3–10 mm long	up to 5 mm long (subsessile), grayish	up to 5 mm long (subsessile)
few, if any, hairs above or below	dense, white, appressed hairs above and below	grayish green above, with spreading hairs; waxy grayish below, with spreading hairs	grayish green above, with flattened hairs; waxy whitish below, with flat hairs
shiny above; some fine hairs at tip of stalk and at base of blade below	sparsely hairy, white to slightly rust-colored above; slightly tan, hairy or glaucous below	grayish green above; grayish green and waxy whitish below, with persistent hairs	grayish green above; grayish green and waxy whitish below, with persistent hairs
prominent, broadly rounded 2–10-mm-long, collar-like stipules at base of leaf stalk; small, warty, toothed glands on stalk at base of blade	stipules minute, soon deciduous	stipules minute, soon deciduous	tiny stipules at base of leaf stalk, soon deciduous
3–12 cm long, dense, appearing with leaves; bracts yellow, with tiny hairs	1–2.5 cm long, subsessile; bracts yellow to light brown or black; stalk short, leafy	4–10 cm long, lax or somewhat limp, silvery green; bracts yellow	4–10 cm long, lax or somewhat limp, silvery green; bracts yellow
2–7 cm long 1–1.5 cm wide	7–15 mm long, stubby, subsessile (stalk 1–2.8 cm long)	2 stamens, hairy near base	2 stamens, hairy near base
shiny orange or yellow, usually hairless (but hairs fine and spreading when present) when young; becoming dark gray, fissured and hairy with age	covered with dense cottony hairs when young; yellow brown-to-blackish, hairless and glossy when old	brown or green when young, with straight, appressed hairs with waxy coating; brownish green, grayish brown and scaly when old, with spreading hairs	brown or green when young, with straight, appressed hairs with waxy coating; smooth, hairless, brownish green, grayish brown and scaly when old

These communities are characterized by aquatic plants that grow entirely submerged or with their leaves floating at the surface. Most of the plants in this chapter root in sand, gravel, silt or mud, but a few are free-floating and drift about with wind and water currents.

Submerged and floating plant communities grow in habitats that range from slow-moving rivers and streams to lakes, ponds, sloughs and pools on floodplains. Water levels usually recede in summer, which often strands some aquatic plants on gravel banks and mud flats but also creates habitat for emergent species.

Extensive communities of single plant species are often found in these habitats. If the water body is large enough, several plant species may be present, each in a discrete zone, with mixtures occurring in the transition areas (ecotones) between zones.

In deep-water marshes, soft-stem and hard-stem bulrush (Scirpus tabernaemontani *and* S. acutus, *chapter 2) often grow far from shore under submerged conditions. Many other aquatic plants are also associated with this habitat, including floating-leaf pondweed* (Potamogeton natans), *Eurasian water-milfoil* (Myriophyllum spicatum), *water smartweed* (Polygonum amphibium, *chapter 2) and water crowfoot* (Ranunculus aquatilis).

In flowing water, submerged masses of Canadian waterweed (*Elodea canadensis*), coontail (*Ceratophyllum demersum*), water crowfoot (*Ranunculus aquatilis*) and curly pondweed (*Potamogeton crispus*), an introduced species, form the most common communities.

In standing water, Mexican water fern (*Azolla mexicana*) and common duckweed (*Lemna minor*) are the most common small-leafed floating species. They often form a distinctive association with greater duckweed (*Spirodela polyrhiza*) and purple-fringed riccia (*Ricciocarpos natans*), an aquatic liverwort. Yellow pond-lily (*Nuphar lutea* ssp. *polysepala*), floating-leaf pondweed (*Potamogeton natans*) and Eurasian water-milfoil (*Myriophyllum spicatum*) form the most

Shallow-water marshes include typically red mats of Mexican water fern (Azolla mexicana) *and combinations of purple-fringed riccia* (Ricciocarpos natans), *duckweed* (Lemna minor *and* Spirodela polyrhiza) *and wapato* (Sagittaria latifolia, *chapter 2).*

Yellow pond-lily (Nuphar lutea *ssp.* polysepala) *can often be seen at the surface of sloughs and lagoons with pondweeds* (Potamogeton *spp.*). *Coontail* (Ceratophyllum demersum) *and Canadian water-weed* (Elodea canadensis) *grow below the surface. Willows* (Salix *spp., chapter 4*) *and reed canary-grass* (Phalaris arundinacea, *chapter 3*) *often grow around slough edges.*

common large-leafed communities in ponds and sloughs.

Howellia (*Howellia aquatilis*) is one of our rarest wetland plants. It typically would grow in shaded, wooded pools and ponds where the water remained clear throughout the season. These pools, commonly associated with the floodplains of streams and rivers, are now usually dominated by a combination of yellow pond-lily and a pondweed, coontail, waterweed or water crowfoot. An association of Oregon ash (*Fraxinus latifolia*, chapter 5), slough sedge (*Carex obnupta*, chapter 4) and snow-berry (*Symphoricarpos albus*, chapter 5) usually surrounds these pools and gradually transitions into upland forest.

The pre-settlement landscape had many thousands of acres of streams and sloughs associated with the floodplains of the Willamette and Columbia rivers and their tributaries. In an effort to improve navigation and control floods on the rivers, the water in many braided streams and sloughs was diverted into a main channel. Many of what were once flowing-water communities have converted to still-water communities, such as those dominated by yellow pond-lily and pondweeds. Many of these 'orphaned channels' have since silt-ed in or were filled or re-contoured for agriculture.

The rare plant howellia (Howellia aquatilis) *is known only from this serene wooded pool, in which it forms a dominant aquatic plant association with yellow pond-lily* (Nuphar lutea *ssp.* polysepala).

MEXICAN WATER FERN
MEXICAN MOSQUITO FERN • WATER VELVET
Azolla mexicana

Salviniaceae (Water fern Family)
INDICATOR STATUS: OBL

GROWTH HABIT: Perennial native water fern; individual plants are commonly up to 1 cm wide, **float in or on water,** often in large mats; **stems are short, flat, floating, hidden by leaves.**

LEAVES: These are most conspicuous feature of plant; **scale-like; overlap one another along stems.**

FLOWERS: Ferns have no flowers but reproduce by spores, which are contained in sporangia.

HABITAT: Ponds, small lakes and backwaters.

NATURAL HISTORY: Mexican water fern is most common from May to August. If it is exposed to nutrient-rich (eutrophic) conditions, water fern can become an undesirable weed. It will choke out other plants from aquatic systems and thereby reduce biological diversity. Anabaena, a filamentous blue-green alga (cyanobacterium), lives in cavities formed by the water fern leaves. The alga fixes nitrogen, which helps provide the water fern with nutrients that are limited in aquatic habitats. Water ferns provide food for dabbling ducks, such as mallards, gadwalls and pintails, as well as teals, shovelers and wigeons.

SIMILAR SPECIES: There are two species of *Azolla* known to the valleys of western Oregon and Washington. Mexican water fern is the most common. Duckweed fern (*A. filiculoides*, OBL) is larger; it averages 2–5 cm long.

NOTES: Mexican water fern is **usually green, but late in the season it often becomes reddish purple**, which gives ponds and lakes a scarlet cast. In places where water fern is cultivated and harvested for use as a fertilizer it is often called 'green manure.'

A: *Mexican water fern* (Azolla mexicana) *in a typical habitat.*
B: *Mexican water fern* (2.3x magnification).

PURPLE-FRINGED RICCIA
Ricciocarpos natans

Ricciaceae (Riccia Family)
INDICATOR STATUS: NOL

Purple-fringed riccia (Ricciocarpos natans, *3.9x magnification*).

GROWTH HABIT: Perennial native floating emergent liverwort; **plants are green with purplish splotches, float on water surface** when not stranded in mud; true **stems are lacking** (as in all liverworts).

LEAVES: Leaf-like plant body (thallus) is flat, 2-lobed, finely dotted with many tiny holes (air cavities); Y-shaped furrows crossing thallus surface give **edges fluted or scalloped** appearance; **lower surfaces are purple**, covered with toothed, flap-like rhizoids, resembling roots, which penetrate and attach to muddy bottoms of ponds and pools.

FLOWERS: Liverworts do not have flowers; they reproduce sexually by spores, or vegetatively by fragmentation.

HABITAT: Ponds, pools, swamps and sloughs.

NATURAL HISTORY: Purple-fringed riccia occurs in early spring and lasts throughout the growing season. It usually grows in warm water or in water that receives nutrient loading. In our area, it almost always grows with common

duckweed (*Lemna minor*), greater duckweed (*Spirodela polyrhiza*) or Mexican water fern (*Azolla mexicana*). Purple-fringed riccia shelters aquatic insects and invertebrates.

SIMILAR SPECIES: The floating leaves of water crowfoot (*Ranunculus aquatilis*, p. 55) can look like purple-fringed riccia, but the presence of stems is a sure sign they are not liverworts. From a distance floating communities of common duckweed (p. 34), greater duckweed (p. 33) or Mexican water fern (p. 31) can be mistaken for purple-fringed riccia. The rhizoids on the undersides of each purple-fringed riccia plant will distinguish it from anything else.

NOTES: Purple-fringed riccia is an aquatic or amphibious liverwort found throughout the tropical regions of the world, including Africa.

GREATER DUCKWEED
Spirodela polyrhiza

Lemnaceae (Duckweed Family)
INDICATOR STATUS: OBL

GROWTH HABIT: Perennial native **floating** forb; each plant consists of small, leaf-like, plant body (thallus) **without stems** but with **5–16 rootlets attached on underside** and extending down into water.

LEAVES: True leaves are lacking; **leaf-like plant body** is **small (4–8 mm long)**, **spoon-shaped**; upper surface is green, with **single red dot**, resists wetting; **underside is purple**.

FLOWERS: Produced within reproductive pouches, very rarely observed in bloom.

HABITAT: Freshwater habitats, such as quiet waters of ponds, lakes and marshes.

NATURAL HISTORY: Greater duckweed is most commonly encountered from June through October. It is common and widely distributed in temperate to tropical regions of the world and is known throughout the United States and southern Canada. In our area, greater duckweed almost always occurs with common duckweed (*Lemna minor*), and together these often form matted 'meadows' on the surface of the water. Although it has not become a pest in our area, it can become a nuisance by creating excessive shading over water. Ducks eat greater duckweed, though they prefer lesser duckweed. Fish are not known to eat it.

SIMILAR SPECIES: Common duckweed (p. 34) is noticeably smaller than greater duckweed and has a single rootlet (see photo).

NOTES: Although still very small, greater duckweed is our largest representative of the duckweed family.

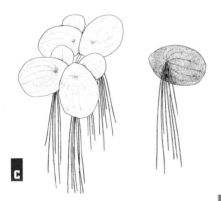

A: *Greater duckweed* (Spirodela polyrhiza, *3.3x magnification*).

B: *Two specimens of greater duckweed* (S. polyrhiza) *on either side of two common duckweed* (Lemna minor) *specimens. Note that greater duckweed has many roots, while common duckweed has only one root.*

C: *Greater duckweed* (S. polyrhiza).

COMMON DUCKWEED
WATER LENTIL • LESSER DUCKWEED
Lemna minor

Lemnaceae (Duckweed Family)
INDICATOR STATUS: OBL

GROWTH HABIT: Perennial native floating forb; individual **plants consist of one minute, leaf-like plant body** (thallus) **with no stem and one underwater rootlet**; colonies form bright green masses on water surface.

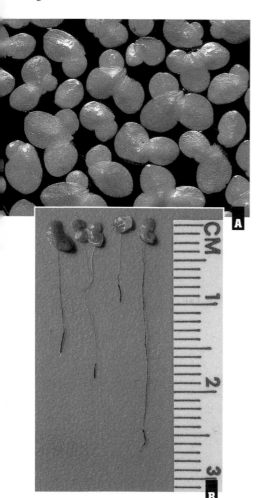

A: *Common duckweed* (Lemna minor, *3.2x magnification).*
B: *Common duckweed.*

LEAVES: True leaves are lacking; leaf-like plant body is 2–5 mm across.

FLOWERS: Borne in fold of plant body; incredibly small, almost never seen, little help in identification.

HABITAT: Quiet backwaters or standing water of shallow or deep marshes, ponds, drainages and small creeks.

NATURAL HISTORY: Common duckweed is most abundant from June through September or later. It usually reproduces by budding, and colonies of genetically identical plants (clonal monocultures) are common. Duckweed can be found in scattered colonies or it can completely blanket small pools and ponds. It may be found stranded in the mud if there has been a drop in the water level. Duckweed is sometimes found growing with cattails (*Typha latifolia*), sedges (*Carex* spp.) and aquatic grasses. The presence of duckweed usually indicates a healthy aquatic environment even though it can give the appearance of stagnation. However, under unhealthy circumstances, such as the influence of too much nitrogen, phosphate or other forms of nutrient loading, it spreads rapidly out of control, choking aquatic systems, and it ultimately produces unfavorable conditions for species diversity and other important wetland functions and values. Waterfowl and shorebirds eat both duckweed and the aquatic insects and invertebrates that live with it.

SIMILAR SPECIES: Star duckweed (*L. trisulca*, OBL), also called ivy duckweed, is very similar, but the individual plants are larger (usually 6–10 mm long) and are connected in branched chains of 10–30 plants. From a distance, common duckweed can be mistaken for other mat-forming aquatic plants, such as algal blooms, greater duckweed (*Spirodela polyrhiza*, p. 33), water ferns (*Azolla* spp., p. 31), purple-fringed riccia (*Ricciocarpos natans*, p. 32) or water-starworts (*Callitriche* spp., p. 60). See also the comparative photos of common duckweed and greater duckweed on page 33.

Lemnaceae (Duckweed Family)
INDICATOR STATUS: OBL

GROWTH HABIT: Perennial native **floating** forb; **tiny, spherical** plants are **stemless, rootless, irregularly lengthened**, 2 mm or less in diameter.

LEAVES: True leaves are lacking; tiny, leaf-like plant body (thallus) is **lady-bug-shaped, grain-like; upper surface is flat with minute, dark brown spots**.

FLOWERS: Minute, occur along edge of plant, rarely seen.

HABITAT: Freshwater marshes, lakes, ponds, pools and sloughs.

NATURAL HISTORY: Watermeal is easily overlooked in aquatic systems because it often intermingles with common duckweed (*Lemna minor*), greater duckweed (*Spirodela polyrhiza*), algae and aquatic liverworts. For this reason, it is probably more widely distributed than is recognized.

SIMILAR SPECIES: Columbia watermeal (*W. columbiana*, OBL), also called Willamette Valley watermeal, is about the size and shape of a pinhead. It is smaller than dotted watermeal and rounder, not as lengthened. See also common duckweed (p. 34) and greater duckweed (p. 44).

NOTES: Watermeals are thought to be the smallest flowering plants in the world. Although the individual plants are barely visible to the naked eye, communities of them form a layer over water that has the consistency of Cream of Wheat, and from a distance they may appear as floating green scum. Dotted watermeal gets its name from the dark spots on the upper surface of each plant.

A

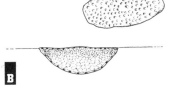

B

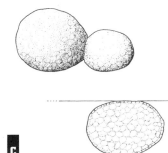

C

A: *The round bodies are Columbia watermeal* (Wolffia columbiana) *and the elongated bodies with white or brownish dots are dotted watermeal* (W. borealis) *(4.3x magnification). The tiny seed pods with gold midsections that appear between the watermeals are of unknown origin. However, since the seeds are large in relation to the the mature watermeals, it is unlikely that either watermeal produced them.*

B: *Dotted watermeal* (W. borealis).

C: *Columbia watermeal* (W. columbiana).

WATER SHIELD · WATER TARGET
Brasenia schreberi

Cabombaceae (Fanwort Family)
INDICATOR STATUS: OBL

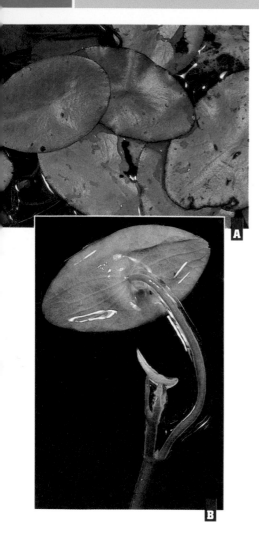

GROWTH HABIT: Perennial native submerged forb; **familiar water-lily appearance**, with large, **round leaves floating on water surface**; stems are slender and slimy, grow from fleshy underground stems (rhizomes) embedded in mud; **submerged parts of plant are covered with thick, gelatinous coating**.

LEAVES: Float neatly on water surface; **disk-like, flat, elliptical**, 5–10 cm across; **smooth-edged (entire)**, without deep notch typical of water-lilies; attached alternately along submerged stems; leaf stalks (petioles) are 5–40 cm long, attached at center of undersides of leaves.

FLOWERS: Borne in leaf axils; relatively small (5–20 cm long), **dull red to purple, with purple stamens**; sometimes open up underwater.

HABITAT: In up to 1.2 m of water in lakes, ponds, sloughs, lagoons and slow-moving streams.

NATURAL HISTORY: Water shield blooms from late July through September. It is most common in coastal lakes and ponds; it is uncommon in the interior valleys of western Oregon and Washington. Water shield often grows in association with pondweeds (*Potamogeton* spp.) and yellow pond-lily (*Nuphar lutea* ssp. *polysepala*) in undisturbed water. Its hard, spherical seeds provide food for waterfowl and muskrats and its floating leaves create shade and shelter for fish.

SIMILAR SPECIES: Water shield can be confused with a water-lily, such as yellow pond-lily (p. 37) and white water-lily (*Nymphaea odorata*, p. 37), but it is easily distinguished from water-lilies by its leaf, which does not have a notch or deep lobe in the blade and has a central attachment of the leaf stalk to the underside of the blade. The leaves of water-lilies have a noticeable notch at the off-center point of attachment of their leaf stalks. Also, only water shield has a gelatinous coating on the undersides of its leaves, and its small, dull, red-to-purplish flowers are quite different from the showy, lotus-like, white and yellow flowers of the water-lilies.

A: *Water shield* (Brasenia schreberi). *Its leaves float on the water surface.*
B: *Water shield has a characteristic gelatinous coating on its stems, buds and leaf undersides. Note also the central attachment of the leaf stalk to the underside of the leaf.*

GROWTH HABIT: Perennial native submerged floating forb; **leaves floating at water surface** are attached by long stalks to underground stems (rhizomes) that are up to 10 cm thick, fleshy and cylindrical, have noticeable leaf scars and lie in muddy bottom sediments.

LEAVES: Floating leaves are large (10–40 cm long) but thin, **broadly oval and heart-shaped, with deep notch at base** where long leaf stalks (petioles) attach; leaf stalks are up to 2.5 m long; leaf edges are often crisped or upturned.

FLOWERS: Borne on flower stalks (peduncles) that **usually extend well above water surface**; **flowers are yellow, bowl-shaped**, with many small, **yellow petals** and smaller, **green, petal-like sepals**; petals and sepals surround central portion, which contains **many purple stamens**.

FRUITS: Fleshy, swollen pods with narrow throat and circular, valve-like opening.

HABITAT: Ponds, lakes, slow-moving streams and deep freshwater marshes.

NATURAL HISTORY: Yellow pond-lily blooms from June through mid-August. It typically needs water 1–2.5 m deep to become established. Once rooted, it often dominates pond and lake communities and may crowd out other plants. Algae, aquatic insects and other invertebrates grow under the leaves. Many wildfowl eat the seeds and some have been known to eat the rhizomes as well. Deer eat the leaves, stems, stalks and flowers, beavers eat the rhizomes, and muskrats eat many parts of the plant.

SIMILAR SPECIES: White water-lily (*Nymphaea odorata*, OBL) is very similar, but it has white flowers and flatter, smaller, rounder leaves. Water shield (*Brasenia schreberi*, p. 36) also has the familiar water-lily appearance, but it does not have a notch in its leaves.

NOTES: The rhizomes of yellow pond-lily were an important food source for Native Americans.

A: *Yellow pond-lily* (Nuphar lutea *ssp.* poly-sepala). *Its large leaves either float on the water surface or grow above it.*

B: *Maturing fruits of yellow pond-lily* (Nuphar lutea *ssp.* polysepala).

C: *White water-lily* (Nymphaea odorata) *is a non-native species that is not as common as yellow pond-lily.*

TAPEGRASS · AMERICAN WILD CELERY
Vallisneria americana

Hydrocharitaceae (Frog's-bit Family)
INDICATOR STATUS: OBL

GROWTH HABIT: Perennial introduced submerged forb; roots are fibrous and produce underground stems (rhizomes) and fleshy propagative buds.

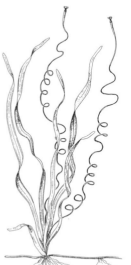

A: *Tapegrass* (Vallisneria americana). *Its grass-like leaves commonly float on the water surface.*

B: *The female flowers of tapegrass are on long, coiled stalks.*

LEAVES: Clustered at stem base; **flattened, ribbon-like, floating**; up to 1 m long, but usually 10–50 cm long and 3–10 mm wide; **spongy, with many lengthwise veins and cross partitions (septa) creating conspicuous, net-like appearance**.

FLOWERS: Female flowers small and white, borne singly at water surface at end of long stalk (peduncle) that tends to coil and can be up to 1–2 m long; several hundred male flowers borne in spherical cluster at end of long stalk; individual male flowers are small (less than 1 mm across), float free when mature.

FRUITS: 5–10 cm long, with **many seeds embedded in gelatinous mass**; exposed when specialized bract (spathe) surrounding female flower retracts and withers after fertilization.

HABITAT: Freshwater habitats, such as quiet waters of ponds, lakes and marshes and backwaters of streams and rivers.

NATURAL HISTORY: Tapegrass is most commonly found from June through September. It is native to the eastern United States. Tapegrass seeds will die if they dry out.

SIMILAR SPECIES: In early spring, with its new leaves, tapegrass can be mistaken for giant bur-reed (*Sparganium eurycarpum*, p. 67), however bur-reed leaves are usually strongly triangular, unlike tapegrass leaves. Other plants with grass- or strap-like leaves, such as cattail (*Typha latifolia*, p. 68) and semi-aquatic mannagrasses (*Glyceria* spp., p. 80), can also be mistaken for tapegrass. Cattail leaves tend to be stiff and erect, however, whereas tapegrass usually has narrower leaves that tend to float on the water surface. Mannagrasses have thinner leaves than tapegrass, and mannagrass leaves lack the net-like pattern found on tapegrass leaves. Once tapegrass plants mature their flowering and fruiting structures easily distinguish them from these other species.

NOTES: This plant was named for an Italian botanist, Antonio Vallisnieri de Vallisnera, (1661–1730), who was especially interested in aquatic plants.

Apiaceae, also called Umbelliferae (Carrot Family)
INDICATOR STATUS: OBL

GROWTH HABIT: Perennial native floating forb; **stems usually trail, creep or float in water**, often spread horizontally.

LEAVES: Kidney-shaped, 1–6 cm wide; deeply 5–6-lobed, appear scalloped; **leathery; bright, shiny green**; leaf stalks (petioles) are 1–5 cm long, delicate.

FLOWERS: In simple, 5–10-flowered umbels that arise from leaf axils on short (1–3 mm long) stalks (peduncles); **flowers are small, yellow, white or greenish**; when in fruit, **stalks curve downwards and eventually tend to curl around stems**.

HABITAT: Marshes, ponds and wet ground along watercourses.

NATURAL HISTORY: Marsh pennywort blooms from July to August. It is an attractive plant, but its floating mats of vegetation can be a nuisance. The seeds and leaves are eaten by wildfowl.

SIMILAR SPECIES: The floating leaves of water crowfoot (*Ranunculus aquatilis*, p. 55) may be mistaken for marsh pennywort.

A: *Marsh pennywort* (Hydrocotyle ranunculoides).

B: *Marsh pennywort has tiny yellow, white or greenish flowers (5x magnification).*

C: *The fruit stalks of marsh pennywort curl down around the creeping stems.*

WATERCRESS
Rorippa nasturtium-aquaticum (Nasturtium officinale)

Brassicaceae, also called Cruciferae (Mustard Family)
INDICATOR STATUS: OBL

Watercress (Rorippa nasturtium-aquaticum).

GROWTH HABIT: Perennial native submerged/emergent forb; stems commonly **floating in water, or prostrate or trailing in mud**; **stems spreading or creeping**, to 20 cm long, **angular**, smooth, crisp, **succulent, weak, easily broken**, often root at lower nodes; exposed roots conspicuous, especially suspended in water, where they appear white and fibrous or thread-like.

LEAVES: **Alternate**; either compound or simple (sometimes both conditions occur on same stem); **compound leaves have 3–9 dark green, succulent, slightly wavy leaflets**; largest leaflet is at tip; other leaflets are in pairs, oval to lance-shaped, up to 4 cm long, slightly separated from each other along main leaf axis (rachis).

FLOWERS: Clustered at end of stalk (peduncle), which can be up to 10 cm tall and lifts flowers well out of water; **flowers are white**, small (less than 1 cm across).

FRUITS: Long pods, about 1–3 cm long and 2 mm wide; curve slightly upwards.

HABITAT: Submerged in shallow, clear, quiet water or in mud along pond edges and marshes; often in artesian springs.

NATURAL HISTORY: Watercress occurs from March to October. It has been introduced into the Pacific Northwest from Europe and is now naturalized through most of the United States. It is easily established and may out-compete other plants in aquatic systems. Watercress tends to form tangled masses that shelter small aquatic life. Muskrats and ducks eat the leaves.

SIMILAR SPECIES: Western bittercress (*Cardamine occidentalis*, FACW+) is similar, but it typically has longer flower stalks (20–40 cm) and larger fruits (2–3 cm long). Willamette Valley bittercress (*C. penduliflora*, p. 105) also has white flowers, but again they are borne at the tips of 20–40 cm long flowering stalks. Yellow marshcress (*R. islandica*, p. 76) is another similar species, but it has yellow flowers.

NOTES: Watercress can be used as a peppery salad green, and contains twice the protein, four times the vitamin C and much more calcium than lettuce.

Potamogetonaceae (Pondweed Family)
INDICATOR STATUS: OBL

GROWTH HABIT: Perennial native floating forb; stems are simple or unbranched, up to 2 mm thick and 1.5 m long; **floating and submerged leaves differ.**

LEAVES: Floating leaves are elliptical, broad (2.5–6.5 cm wide), 5–10 cm long, **brownish green or copper colored; very long leaf stalks (petioles)**; abrupt joint in leaf stalk where it joins leaf allows blade to float flat on water surface; **submerged leaves are leathery, long (10–20 cm), narrow (1–2 mm wide).**

FLOWERS: In 10–12 whorls densely clustered along cylindrical **spikes that reach 5 cm in length when fruit mature.**

HABITAT: Shallow water of ponds, pools and lakes or quiet backwaters of streams.

NATURAL HISTORY: Floating-leaf pondweed is most commonly found from May through August. It is very common on both sides of the Cascades and can be an indicator of brackish water. The underground stems (rhizomes) and achene fruits of floating-leaf pondweed remain late in the season and are good food stores for ducks.

SIMILAR SPECIES: Floating-leaf pondweed can be distinguished from other pondweeds by its **extremely narrow (usually less than 2 mm wide), leathery submerged leaves.**

A

A, B: *Floating-leaf pondweed* (Potamogeton natans) *is one of the most common pondweeds in the Pacific Northwest.*

B

GRASSY PONDWEED
VARIABLE PONDWEED • GRASS-LEAF PONDWEED
Potamogeton gramineus

Potamogetonaceae (Pondweed Family)
INDICATOR STATUS: OBL

A, B: *Grassy pondweed* (Potamogeton
gramineus). *Note the short, oval floating
leaves and the long, linear submerged
leaves.*

GROWTH HABIT: Perennial native submerged
forb; stems are round and slender; **floating and
submerged leaves differ**.

LEAVES: In two main forms, with intermediate
forms found near top of plant; **floating leaves
are small, wide, elliptical, 2–5 cm long, 1–2 cm
wide**; attached to main stem on **long leaf stalks
(petioles); submerged leaves are 3–9 cm long,
narrow** (3–10 mm wide); **attached directly to
stems (sessile)**; conspicuous veins run along
their length.

FLOWERS: In erect spikes that extend well
above water surface; flowers are small, inconspicuous.

HABITAT: Lakes and slow-moving streams.

NATURAL HISTORY: Grassy pondweed is most
abundant in valley habitats from June through
August. It occurs on both sides of the Cascades
and is a common pondweed in shallow water.
The tubers and other parts of grassy pondweed
are a choice food for shorebirds, ducks and
geese.

SIMILAR SPECIES: Several other pondweeds,
especially floating-leaf pondweed (*P. natans*,
p. 41), ribbon-leaf pondweed (*P. epihydrus*,
p. 43), large-leaf pondweed (*P. amplifolius*,
OBL) and long-leaf pondweed (*P. nodosus*,
OBL), also have two types of leaves. The floating leaves of all these species are more or less
elliptical-lance-shaped to oblanceolate, and
their submerged leaves tend to be longer and
narrower. The floating leaves of both large-leaf
pondweed and long-leaf pondweed are generally larger than those of grassy pondweed. In
large-leaf pondweed the floating leaves are
5–10 cm long and 2–4 cm wide; in long-leaf
pondweed the floating leaves are 5–12 cm long
and 2–4 cm wide. Grassy pondweed can be distinguished from all these other pondweeds with
two leaf forms by the fact that its floating leaves
are not leathery. Grassy pondweed is more
uncommon than these other species.

Potamogetonaceae (Pondweed Family)
INDICATOR STATUS: OBL

GROWTH HABIT: Perennial native floating forb; each plant has one flat main stem; **stems and leaves have overall slippery and flattened character**; floating and submerged leaves differ.

LEAVES: Floating leaves are firm, **oval**, 4–8 cm long, 1–2 cm wide; appear to attach directly to main stem, but actually have short, flattened leaf stalks (petioles); submerged leaves are **ribbon-like, 10–20 cm long, 3–10 mm wide, slippery, limp (flaccid)**; disintegrate quickly out of water; attached directly to stems without stalks; 5–7 parallel veins.

FLOWERS: Densely packed in cylindrical spikes; flowers are brownish green.

HABITAT: Lakes, ponds, pools, streams and quiet shallow and deep water; especially common near inlets and outlets of spring-fed ponds.

NATURAL HISTORY: Ribbon-leaf pondweed is found mostly in June and July. It occurs on both sides of the Cascades.

SIMILAR SPECIES: Other pondweeds, especially floating-leaf pondweed (*P. natans*, p. 41), look very similar. The floating leaves of floating-leaf pondweed are almost indistinguishable from those of ribbon-leaf pondweed, but the submerged leaves of these two species differ. The submerged leaves of floating-leaf pondweed are noticeably leathery and are much narrower than those of ribbon-leaf pondweed, which are wider and ribbon-like. It can be very difficult to distinguish between the different pondweed species, especially since they often cross-breed and form hybrids with intermediate characteristics.

A: *Ribbon-leaf pondweed* (Potamogeton epihydrus).
B: *Ribbon-leaf pondweed in a typical habitat.*

WHITE-STALKED PONDWEED • LONG-STALKED PONDWEED • WHITE-STEMMED PONDWEED
Potamogeton praelongus

Potamogetonaceae (Pondweed Family)
INDICATOR STATUS: OBL

White-stalked pondweed (Potamogeton praelongus).

GROWTH HABIT: Perennial native floating forb; **stems are unbranched or infrequently branched, 2–3 mm thick, 2–3 m long,** usually **yellowish green,** sometimes **whitish;** shorter nodes between leaves commonly zigzag.

LEAVES: All are submerged, usually under deep water; **oblong, lance-shaped, 10–25 cm long, 2–3 cm broad;** attached directly to stem without stalks (**sessile**); 3–5 prominent veins run length of blade, with less obvious cross veins; leaf tip is rounded, often hooded.

FLOWERS: In 6–12 whorls densely clustered along cylindrical spikes that reach 2.5–5 cm in length when fruit mature; main axis (rachis) of spike can be shaped like club or baseball bat.

HABITAT: Deep water of ponds, lakes and streams.

NATURAL HISTORY: White-stalked pondweed is most commonly found from May through July. It is found on both sides of the Cascades, most often in the deep-water habitats of streams and lakes in the southern valleys. Ducks eat the underground stems (rhizomes) and fruits of white-stalked pondweed.

SIMILAR SPECIES: The white stems of white-stalked pondweed usually distinguish it from the other pondweeds.

Potamogetonaceae (Pondweed Family)
INDICATOR STATUS: OBL

Curly pondweed (Potamogeton crispus).

GROWTH HABIT: Perennial introduced submerged forb; **plants are slippery**; **stems are flattened** and somewhat branching, 40–80 cm long, mostly 1–2 mm wide.

LEAVES: Attached directly to stem (sessile); long, narrow, **crisped along edges**; delicate bracts at leaf bases tend to shred when pulled apart.

FLOWERS: In cylindrical or head-like spike that arises from leaf axils on long stalk (peduncle); flowers are brownish and inconspicuous.

HABITAT: Lakes, ponds, pools and slow-moving water of streams and rivers.

NATURAL HISTORY: Curly pondweed is most commonly found from late June through August. It was introduced from Europe and is now well established in lakes and streams. Curly pondweed tends to increase oxygen levels and produce substantial organic material in aquatic environments, but it sometimes becomes a pest in waterways, lakes and reservoirs. This pondweed shelters small fish and aquatic insects that provide food for larger fish and amphibians.

SIMILAR SPECIES: A few of the pondweeds are difficult to differentiate from one another. Fortunately curly pondweed had a very distinct leaf type unlike most anything else growing in water.

LEAFY PONDWEED
Potamogeton foliosus

Potamogetonaceae (Pondweed Family)
INDICATOR STATUS: OBL

A

GROWTH HABIT: Perennial native submerged forb; **stems are extremely thin, flattened and sparingly branched**.

LEAVES: All submerged; **bright green, very narrow** (1–2 mm wide), to 10 cm long; smaller leaf-like stipules closely surround stems.

FLOWERS: In 2–3 whorls in compact spikes (barely 5 mm long) on short (rarely over 2 cm long), stout stalks (peduncles); individual flowers are tiny.

HABITAT: Standing or slow-moving, shallow water of pools, ponds, lakes and reservoirs.

NATURAL HISTORY: Leafy pondweed appears in the Northwest from late June through August. It occurs on both sides of the Cascades and throughout most of the United States, including Hawaii, and Mexico. Leafy pondweed is often found in brackish, shallow water and it is often associated with filamentous algae, such as brittlewort (*Nitella* sp.) and stonewort (*Chara* sp.).

SIMILAR SPECIES: Sago pondweed (*P. pectinatus*, OBL) is more freely and dichoto-

mously branched than leafy pondweed and has a longer (up to 10 cm long), more delicate flower stalk. Horned pondweed (*Zannichellia palustris*, OBL) has more slender and delicate stems and leaves. Water-nymphs (*Najas* spp., p. 48) have leaves that are uniform in length, and although they appear to be in whorls of four around the stem, the leaves are actually opposite. Leafy pondweed can also be confused with brittlewort or stonewort algae, which often grow with it. Stonewort is gray-green and gritty (due to lime deposits) and has a skunk-like odor. Brittlewort is more delicate and greener than stonewort, and does not have a disagreeable odor. Brittlewort and stonewort have groups

B

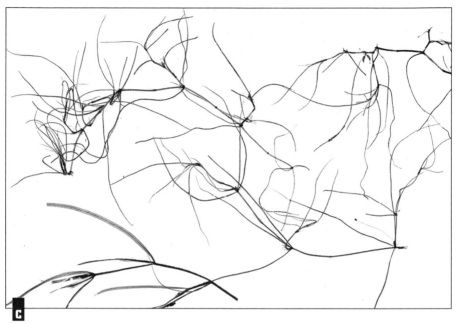

C

of 5–7 whorled 'leaves' along their stems, compared to the more branched leaf arrangement of leafy pondweed.

NOTES: Despite its name, leafy pondweed, at least specimens occurring in the Pacific Northwest, does not usually have very many leaves.

D

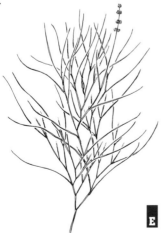

E

A, B: *Leafy pondweed* (Potamogeton foliosus).
C: *Horned pondweed* (Zannichellia palustris).
D: *Leafy pondweed* (P. foliosus), *showing a close-up of the leaves and fruits (2x magnification).*
E: *Sago pondweed* (P. pectinatus).

GUADALUPE WATER-NYMPH
COMMON WATER-NYMPH • SOUTHERN NAIAD
Najas guadalupensis

Hydrocharitaceae (Frog's-bit Family)
INDICATOR STATUS: OBL

A

B

A: *Guadalupe water-nymph* (Najas guadalupensis).
B: *Wavy water-nymph* (N. flexilis), *showing detail of a leaf and seed.*

GROWTH HABIT: Annual native submerged forb; **delicate**; attaches to bottom sediments and **floats just below water surface**; **stems are slender, limp** (flaccid), branched, grow to approximately 60 cm long.

LEAVES: Appear to be in whorls of four around stem, but actually **in opposite pairs**; **thread-like**, evenly spaced, **glossy green**, mostly submerged; **leaf bases are broad and clasp stem**.

FLOWERS: In clusters at base of broadened leaf stalk (petiole); tiny, fairly well concealed.

HABITAT: Shallow, quiet water of ponds, lakes, pools and slow-moving streams.

NATURAL HISTORY: Guadalupe water-nymph is most often found from June through August. It sometimes forms mats on the water surface, but it is usually well submerged. It is a good source of food and shelter for fish. The stems, leaves and seeds are a choice food for wildfowl, marsh birds, ducks and muskrats.

SIMILAR SPECIES: Wavy water-nymph (*N. flexilis*, OBL) is very similar, but it has narrow, long-tapering leaves, while Guadalupe water-nymph has shorter, blunt-tipped leaves. The submerged leaves of water crowfoot (*Ranunculus aquatilis*, p. 55) are finely thread-like, like water-nymph leaves, but water crowfoot is easily distinguished by its larger, lobed floating leaves. The water-nymphs' forked stems and broadened leaf bases distinguish them from water-starworts (*Callitriche* spp., p. 60), which have broader leaves in an opposite arrangement. Water-nymphs can also be mistaken for two types of algae—brittlewort (*Nitella* sp.) and stonewort (*Chara* sp.)—which have their 'leaves' in whorls of 5–7 along the stem. Stonewort is gray-green and gritty (due to lime deposits) and has a skunk-like odor. Brittlewort is more delicate and greener, and does not have a strong odor.

NOTES: The water-nymph genus name derives from the Greek *naias*, 'naiad,' the name for the nymphs that inhabit springs and streams in Greek mythology.

GROWTH HABIT: Annual native submerged forb; **stems are long** (10–60 cm), **hairless (glabrous), green, thin and delicate, branch out at ends,** easily broken.

LEAVES: Most are alternate, or they appear opposite or in whorls of three along stems; limp (flaccid), thread-like, up to 1.5 mm wide (barely wider than stems).

FLOWERS: In leaf axils along stems; flowers are **tubular,** large relative to stems and leaves; petals are often absent; **five sepals,** 1.5–7 mm long; bracts are stouter and broader than leaves and have shallowly lobed tips.

FRUITS: As fruit matures, **sepals surround top of capsule and become conspicuous part of it;** capsule swells and lengthens as it matures; seeds are shiny brown.

HABITAT: Bottoms of ponds or small lakes.

NATURAL HISTORY: Howellia occurs from May through July, but it is one of our rarest wetland plants. It was first discovered in Oregon, but it has not been seen in that state since the 1970s. It was also reported from California but has not been seen there for many years. It is now known only from a few sites in Washington, Idaho and Montana. Howellia often becomes stranded in the mud along the edges of pools, ponds and small lakes as water levels drop. Its stem and leaves are home to small aquatic insects and egg masses.

SIMILAR SPECIES: Water-nymphs (*Najas* spp., p. 48) are denser, more compact plants with fairly uniform, dichotomous branching, while howellia is more trailing once it is detangled from its usual, stringy mass. Also, howellia has distinctive, tubular flowers on stout flower

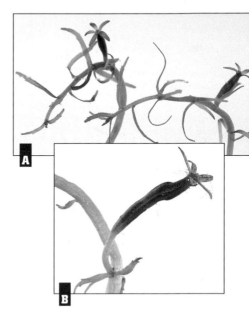

stalks (peduncles), while water-nymph flowers are hidden in the leaf axils. Howellia can also be mistaken for two types of algae—brittlewort (*Nitella* sp.) and stonewort (*Chara* sp.). Stonewort is gray-green and gritty (due to lime deposits) and has a skunk-like odor. Brittlewort is more delicate and greener, and does not have a disagreeable odor. Like howellia, these algae have groups of 5–7 whorled 'leaves' along their stems. The leaves of all three species grow to about the same size, but in the algae, one of the 'leaves' of the whorl grows into a new branch that gives rise to more whorls of 'leaves.'

NOTES: Howellia was named for Thomas (1842–1912) and Joseph (1830–1912) Howell, who were among the earliest resident botanists of the Pacific Northwest.

A: *Howellia* (Howellia aquatilis, *2.3x magnification). Most leaves are long and linear, but the bracts are shorter and stouter, with subtle notches around their edges.*

B: *Detail of howellia's fruiting structure (4x magnification).*

C: *Howellia.*

CANADIAN WATERWEED · DITCHMOSS
Elodea canadensis

Hydrocharitaceae (Frog's-bit Family)
INDICATOR STATUS: OBL

A

B

GROWTH HABIT: Perennial native submerged forb; plants often **free-floating**, even when they appear to ascend from branching stems; short roots usually grow from regularly spaced nodes along stems, but usually not attached to substrate.

LEAVES: Usually grouped **in whorls of three, opposite towards stem base**; **narrow** (6–15 mm long, 1.5–2.5 mm wide), **taper abruptly to fine point**; **leaf edges are slightly toothed**.

FLOWERS: Float on water surface; small, white or pinkish.

HABITAT: Ponds, lakes, lagoons, sloughs and slow-moving streams.

NATURAL HISTORY: Canadian waterweed, a native species, is most abundant from July through September. Waterweeds rarely produce seeds. Instead they reproduce by fragmentation, and drifting pieces of waterweed are common in water where it grows. Waterweed provides shelter for small fish and for many insects and small invertebrates that are important prey for larger fish, frogs and salamanders. The leafy stems are eaten by waterfowl (generally ducks), beavers and muskrats.

SIMILAR SPECIES: South American waterweed (*Egeria densa*, also called *Elodea densa*, OBL) is an introduced pest that often gets out of control, especially in nutrient-rich (eutrophic) waters. It prefers somewhat brackish or alkaline water and is more common in the southern Willamette Valley and in coastal lakes. It has larger leaves that are usually 2.5–3 cm long and 2–3 mm wide.

NOTES: Waterweeds are popular aquarium plants because they increase water oxygen levels and thrive in artificial habitats. All species in the frog's-bit family (Hydrocharitaceae) are aquatic.

A: *Canadian waterweed* (Elodea canadensis) *is native to the Pacific Northwest.*

B: *South American waterweed* (Egeria densa), *a common aquarium plant, is a non-native pest in lakes and ponds.*

C: *The leaves of Canadian waterweed* (Elodea canadensis) *are narrow.*

D: *The leaves of South American waterweed* (Egeria densa) *are wider.*

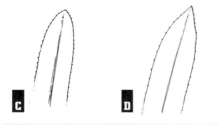

C   **D**

Ceratophyllaceae (Hornwort Family)
INDICATOR STATUS: OBL

GROWTH HABIT: Perennial native **submerged forb**; bushy plant, resembles a tail because of the **dense, bulky whorls of leaves**; stems are long, stiff, brittle, with many branches, may be embedded in bottom sediments or **free-floating** below water surface; rootless.

LEAVES: In dense, bulky whorls around stems; usually 5–12 leaves per whorl; **each leaf forks into thread-like (filiform) segments** about 0.2–0.8 mm wide; fine teeth (commonly thought of as horns) along one side of each segment.

FLOWERS: Borne in leaf axils; inconspicuous; flowers (and fruits) show as tiny, red cylinders.

HABITAT: Ponds, lakes, sloughs and slow-moving streams.

NATURAL HISTORY: Coontail is most abundant from June through September. It can form extensive beds and is used for food and cover by birds, muskrats and fish. The seeds are eaten by mallards, gadwalls and shorebirds, but generally as a second or third choice. Coontail increases oxygen levels where it grows.

SIMILAR SPECIES: Coontail can be mistaken for one of the water-milfoils (*Myriophyllum* spp., p. 52). The submerged leaves of water crowfoot (*Ranunculus aquatilis*, p. 55) are finely thread-like and resemble coontail leaves, but water crowfoot usually has two leaf types, and the floating leaves are much larger (5–15 mm long and twice as broad) and they are both shallowly and deeply 3-lobed. Waterweeds (*Elodea canadensis* and *Egeria densa*, p. 50) have much broader leaves, generally 6–15 mm long and 1.5–2.5 mm wide. Bladderworts (*Utricularia* spp., p. 54) are easily distinguished by the distinct, clear or blackened bladders along their stems. Water-nymphs (*Najas* spp., p. 48) have long-tapering or blunt-tipped leaves with broadened leaf bases. Also, although water-nymph leaves appear to be in whorls, they are actually opposite.

NOTES: Coontail is a bushy-looking plant that is commonly grown in aquaria. *Ceratophyllum* is one of the few genera of vascular plants that do not have roots.

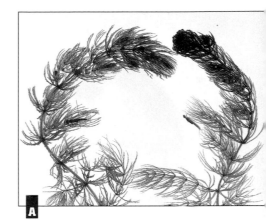

A

B

A, B: *Coontail (Ceratophyllum demersum). (Note: a damsel-fly larva is attached to the coontail in the photograph.)*

WATER-MILFOILS
Myriophyllum spp.

Haloragaceae (Water-milfoil Family)
INDICATOR STATUS: OBL

A

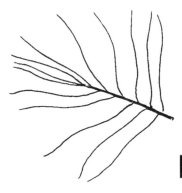

B

C

GROWTH HABIT: Perennial native/introduced submerged forbs; stems green to light-brown or reddish, simple or branched, grow to 2–3 m long.

LEAVES: Parrot's feather (*M. aquaticum*, OBL) leaves are usually **divided into 13–30 thread-like (filiform) segments** that appear **feather-like** and are grouped in 3–6 whorls around stem; mostly submerged; light green, olive or brown. Western water-milfoil (*M. hippuroides*, OBL) leaves are very similar, but they are usually more than 2 cm long.

FLOWERS: Borne along stem above water; a bract-like scale occurs below flowering structure.

HABITAT: Root in sandy, silty bottoms of lakes, ponds, reservoirs and marshes.

NATURAL HISTORY: Water-milfoils are present year round. Waterfowl and shorebirds occasionally eat the fruits and foliage of these plants, and they are also eaten sparingly by muskrats. Western water-milfoil provides shelter for fish and supports many aquatic insects and other invertebrates. It is the only water-milfoil native to North America, and it is not considered a pest. Parrot's feather and Eurasian water-milfoil (*M. spicatum*, OBL), which were introduced to North America from Europe, are aggressive aquatic weeds. Efforts are underway to discourage their growth in lakes and reservoirs, where they form dense mats that obstruct swimmers, boaters, and even fish. Eurasian water-milfoil tolerates brackish water.

SIMILAR SPECIES: Coontail (*Ceratophyllum demersum*, p. 51) is similar, but its leaves are arranged in whorls that branch out in a forked rather than feather arrangement. The submerged leaves of water crowfoot (*Ranunculus aquatilis*, p. 55) are deeply divided like those of parrot's feather, but the floating leaves of water crowfoot are broadly lobed, distinguishing it from the water-milfoils. Common bladderwort (*Utricularia macrorhiza*, p. 54) is distinguished by the clear or blackened bladders along its stems. Canadian waterweed (*Elodea canadensis*), with its shorter, less dense leaves, can be confused with western water-milfoil. These two species have similar leaves emerging from the water, but the submerged leaves of western water-milfoil are divided many times, while the submerged leaves of Canadian waterweed are not divided.

NOTES: Parrot's feather is also known as *M. brasiliense*. The three or more species of *Myriophyllum* that occur in aquatic habitats in our region are sometimes difficult to distinguish from one another. Many have different forms of foliage on the same plants, an adaptation to aquatic and emergent environments.

A: *Western water-milfoil* (Myriophyllum hippuroides).

B: *The tiny flowers of western water-milfoil* (M. hippuroides) *are borne in the leaf axils (2.5x magnification).*

C: *Western water-milfoil* (M. hippuroides).

D: *Parrot's feather* (M. aquaticum) *is a non-native pest.*

E: *Parrot's feather* (M. aquaticum).

COMMON BLADDERWORT
Utricularia macrorhiza (U. vulgaris)

Lentibulariaceae (Bladderwort Family)
INDICATOR STATUS: OBL

GROWTH HABIT: Perennial native submerged forb; aquatic, insectivorous plant with unique, **paired, bladder-like cavities** attached at leaf axils along stems; rootless, free-floating or sometimes attached to substrate; stems are slender, 30–300 cm long.

LEAVES: Fine, light green, **divided many times into hair-like segments**.

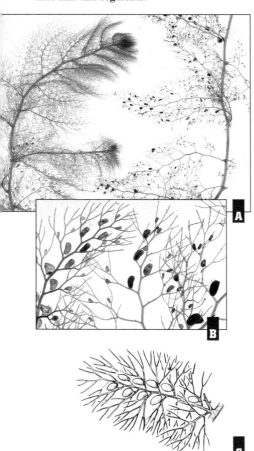

FLOWERS: Borne well above water on upright flower stalk (scape); flowers are fairly large, **2-lipped, yellow**, rarely seen.

HABITAT: Still or slow-moving waters of ponds, lakes and streams.

NATURAL HISTORY: The most distinctive feature of common bladderwort is its bladders. When a small crustacean or other aquatic invertebrate brushes against a bladder it trips a trigger-hair near the bladder's valve-like opening or 'trapdoor.' Triggering the watertight bladder creates a slight inrush of water that sucks in the invertebrate. The opening then shuts and the bladder fills with enzymes that digest the prey. The bladders are clear green at first, but they become purple and then black with age as animal remains collect and decay inside them. Bladderwort can become weedy, especially when it grows with water-milfoils (*Myriophyllum* spp.) and coontail (*Ceratophyllum demersum*), which it often does. These communities often crowd and choke waterways, lakes and reservoirs, reducing biological diversity. Bladderwort provides good food and cover for fish, which feed on the small aquatic animals and algae harbored among its stringy mats (at least those animals not captured by the bladderwort). Bladderwort is eaten by muskrats.

SIMILAR SPECIES: Mountain bladderwort (*U. intermedia*, OBL), which is rare in Washington, lesser bladderwort (*U. minor*, OBL), which is listed as rare in Oregon, and humped bladderwort (*U. gibba*, OBL), which is also rare, are three other species in our region. All three have generally smaller flowers and leaves than common bladderwort. Mountain bladderwort is also distinguished by having its bladders on special, leafless stems. Other aquatic species, including Guadalupe water-nymph (*Najas guadalupensis*, p. 48), water-milfoils (p. 52), coontail (p. 51), waterweeds (*Elodea canadensis* and *Egeria densa*, p. 50), water-starworts (*Callitriche* spp., p. 60), and water crowfoot (*Ranunculus aquatilis*, p. 55) can also be mistaken for common bladderwort, but a close inspection of the stems for the bladders should prevent any confusion (consult technical manuals).

A: *Common bladderwort* (U. macrorhiza).
B: *The insectivorous bladders of common bladderwort become blackened with the remains of invertebrates (1.8x magnification).*
C: *Common bladderwort.*

Ranunculaceae (Buttercup Family)
INDICATOR STATUS: OBL

GROWTH HABIT: Perennial native submerged forb; **stems are 20–60 cm long, tend to branch, trail and entwine underwater**; roots growing from lower nodes further entangle submerged parts of plant; **floating and submerged leaves differ in form and shape**.

LEAVES: Submerged leaves are finely divided into many thread-like segments; floating leaves are 5–15 mm long, 1–3 cm broad, **deeply palmately lobed into three distinct segments** that are again 3–5-lobed, though much less deeply, **appears scalloped**.

FLOWERS: Small, white; five petals; many yellow stamens give appearance of yellow centers.

HABITAT: Ponds, pools, ditches, irrigation canals and sluggish streams.

NATURAL HISTORY: Water crowfoot blooms from May until August. It is common in floodplains and valleys of the Pacific Northwest, but also occurs at higher elevations on both sides of the Cascades. Water crowfoot plants usually have many more thread-like, submerged leaves than broad, floating leaves, especially in early spring. Later in the growing season, when water crowfoot plants are often in shallow water, the floating leaves become more numerous, though there are still less of them than the submerged leaves. Water crowfoot can become very large and dense in lakes.

SIMILAR SPECIES: Lobb's water buttercup (*R. lobbii*, OBL) looks much like water crowfoot, but the floating leaves of Lobb's water buttercup

are smaller (0.5–1 cm long and 2.5–5 mm wide), and often have purplish splotches. Lobb's water buttercup mainly grows in coastal regions. Yellow water buttercup (*R. flabellaris*, OBL) is also aquatic and has two leaf forms. It has yellow flowers and grows along the Columbia Gorge, mostly east of the Cascades. Because of their highly divided submerged leaves, water buttercups can be mistaken for water-milfoils (*Myriophyllum* spp., p. 52), coontail (*Ceratophyllum demersum*, p. 51), Guadalupe water-nymph (*Najas guadalupensis*, p. 48), Canadian waterweed (*Elodea canadensis*, p. 50), and common bladderwort (*Utricularia macrorhiza*, p. 54).

A: *Water crowfoot* (Ranunculus aquatilis).
  Note the highly dissected and thread-like submerged leaves.

B: *A tangled mass of water crowfoot in a typical habitat.*

C: *A submerged leaf of water crowfoot.*

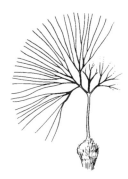

WATER MOSS · BROOK MOSS
Fontinalis antipyretica

Fontinalaceae (Water moss Family)
INDICATOR STATUS: NOL

A

GROWTH HABIT: Perennial native **submerged moss**; **dark green**; **stems are entirely covered by leaves.**

LEAVES: Arranged **in three rows (3-ranked)**, closely **overlapping, all oriented upwards**; rather rigid, concave, **keel-shaped.**

FLOWERS: Mosses reproduce by spores that develop on reproductive structures called sporophytes; water moss sporophytes are almost hidden by overlapping leaves.

HABITAT: Rocks and wood in ponds, lakes, lagoons, sloughs and slow-moving streams.

NATURAL HISTORY: Water moss is most abundant from July through September. Its overlapping, trough-like leaves are home to aquatic insects, larvae and other microscopic life. Water moss is also an important nesting and breeding habitat for Oregon chub, an endangered species of minnow found only in the Willamette River and its tributaries.

SIMILAR SPECIES: Water moss is quite distinctive, but it may be confused with waterweeds (*Elodea* spp., p. 50).

NOTES: Water moss is a true moss, and like all species of its family (Fontinalaceae) it is aquatic. It rehydrates well after drying out.

A: *Water moss* (Fontinalis antipyretica). *Note the spore capsules near the middle of the photograph.*
B: *Water moss leaves overlap one another and hide the stems.*

B

These communities, which include marshes in a variety of land forms, all have water at the surface for most or all of the growing season and are characterized by emergent vegetation. They occur along the edges of rivers, streams, sloughs, lakes and ponds, as well as in wet prairies. Freshwater tidal communities on the lower Columbia and Willamette rivers are also included.

Water levels typically recede in summer, exposing gravel or mud flats and creating habitat for some seasonal species. Aquatic species may become stranded in these communities and survive, often with modified forms, if the substrate is wet enough. Sedges, spike-rushes, rushes and certain grass species are sometimes more abundant in these communities than broad-leaved plants.

Soft-stem bulrush (Scirpus tabernaemontani) *and cattail* (Typha latifolia) *are the dominant plants of a typical shoreline community.*

Marshy shore communities are usually divided into distinct zones based on differences in water level, from the deepest levels of the system to the gradually shallower levels moving towards the shoreline. Weak-stemmed or reclining plants grow in the deeper water and are often specifically adapted to occasional exposure on mud flats. The taller herbs and shrubs grow on the relatively drier parts of the shore.

Knotgrass (Paspalum distichum).

Perennially wet marshes are usually too wet for reed canary-grass (*Phalaris arundinacea*, chapter 3), a noxious weed, to become established, and native species, such as creeping spike-rush (*Eleocharis macrostachya*), cattail (*Typha latifolia*), small-fruited bulrush (*Scirpus microcarpus*),

Mud flats along the floodplains of major rivers with tidal influence typically have communities of soft-stem bulrush (Scirpus tabernaemontani), *wapato* (Sagittaria latifolia), *awned flat-sedge* (Cyperus aristatus) *and knotgrass* (Paspalum distichum). *Pictured here is one plant community type of this habitat.*

Nodding beggarticks (Bidens cernua) *and water-pepper* (Polygonum hydropiperoides) *form a common association in marshy shore habitats. False loosestrife* (Ludwigia palustris)*, which is not shown here, is also found in this community.*

Awned flat-sedge (Cyperus aristatus).

soft-stem bulrush (*Scirpus tabernaemontani*) and giant bur-reed (*Sparganium eurycarpum*), often form monotypic communities.

False loosestrife (*Ludwigia palustris*) and waterpepper (*Polygonum hydropiperoides*) make up a common community on floodplains of the lower Willamette and Columbia rivers, in areas that flood in spring and dry down by early summer. Water smartweed (*Polygonum amphibium*) also occurs occasionally as a community type in the same region, as well as throughout the range.

Knotgrass (*Paspalum distichum*) forms an uncommon community type in perennially flooded marshes along the lower Columbia River, and may be a relict of the pre-settlement vegetation, which has been eliminated from much of its former range.

Channel dredging, flood control and agriculture have caused extensive losses of these marshy habitats. Wapato (*Sagittaria latifolia*) communities were once common throughout our region in floodplain marshes that remained flooded until midsummer. Flood control has cut off the seasonal recharge of these marshes, and wapato has been displaced by reed canary-grass and siltation. Wapato is now restricted primarily to areas of daily tidal fluctuation along the Columbia River, where flooding excludes reed canary-grass. Spike-rush marshes along the lower Columbia River, with interesting mud-flat species, such as awned flat-sedge (*Cyperus aristatus*), mudwort (*Limosella aquatica*), pygmy-weed (*Tillaea aquatica*), horned pondweed (*Zannichellia palustris*, chapter 1) and mud-purslane (*Elatine* sp.), have suffered a similar fate.

Receding shoreline communities often contain wapato (Sagittaria latifolia) *and floating-leaf pondweed* (Potamogeton natans)*, which can tolerate being stranded in the mud when water levels drop.*

Ranunculaceae (Buttercup Family)
INDICATOR STATUS: FACW

GROWTH HABIT: Perennial native emergent forb; **delicate**, semi-aquatic plant with **bright yellow buttercup flowers**; single stem or main stem that branches near top and roots at lower nodes; **stems are smooth, hairless, 10–50 cm long, either float on water surface or lay prostrate and creeping along ground**.

LEAVES: Simple; smooth-edged (entire); **narrowly linear to oblanceolate, 2–6 cm long**; attached in opposite pairs; upper leaves usually have short (2–8 cm long) leaf stalks (petioles) or appear to attach directly to main stems; lower leaves have long leaf stalks.

FLOWERS: 2–25 per plant, on long (up to 10 cm), slender, often lax stalks (pedicels) from main stems; flowers are glossy, bright yellow, 1–1.5 cm across; petals are 3–6 mm long.

FRUITS: Egg-shaped, hairless achenes with short, curved beaks.

HABITAT: Mud or water bordering lakes, ponds and streams, and wet meadows, prairies and wooded wetlands where there is standing water.

NATURAL HISTORY: Creeping spearwort blooms in June and July. It tolerates brackish water well and grows in many areas of the world.

SIMILAR SPECIES: Two other buttercups found in similar habitats are celery-leaf buttercup (*R. sceleratus*, OBL) and yellow water buttercup (*R. flabellaris*, OBL). Celery-leaf buttercup leaves are 3–5-lobed and are scalloped around their edges. Yellow water buttercup is an aquatic species that also has yellow flowers, but it has two leaf forms. Its submerged leaves are divided into very narrow, thread-like segments. Yellow water buttercup grows along the Columbia River Gorge and mostly east of the Cascades.

A: *Creeping spearwort* (Ranunculus flammula) *is found in shady areas around pond and lake edges and in very wet woods.*

B: *Celery-leaf buttercup* (R. sceleratus) *is found in similar habitats, especially along the Columbia River.*

A

B

GROWTH HABIT: Annual native/introduced submerged floating forbs; usually in **floating, tangled mats with thread-like stems and clusters of green leaves**; stems are delicate and limp, grow supported by water or stranded in mud.

LEAVES: Attached in **opposite pairs**; upper leaves float on water surface in clusters or rosettes, individual leaves mostly spoon-shaped, usually 5–8 mm wide (but can be 1 cm wide) and have 1–3 veins; lower leaves submerged, linear in some species, 5–25 mm long.

FLOWERS: In groups of 1–3 in leaf axils; flowers are tiny and inconspicuous; a pair of bracts occurs below each flower.

FRUITS: Tiny, rounded achenes.

HABITAT: Lakes, ponds, ditches, sluggish streams and other quiet waters.

NATURAL HISTORY: Water-starworts appear **in early spring** and bloom from May to August. When water levels drop, water-starworts survive stranded on muddy shores or embankments. These amphibious plants are eaten by waterfowl. They often form mats that shelter small aquatic animals.

SIMILAR SPECIES: Several species of water-starwort are found at low elevations and on floodplains in the Pacific Northwest. Four of the most common are different-leaf water-starwort (*C. heterophylla*, OBL), pond water-starwort (*C. stagnalis*, OBL), an introduced species, vernal water-starwort (*C. verna*, OBL) and autumnal water-starwort (*C. hermaphroditica*, OBL). These species can be difficult to distinguish from one another. The leaf shapes of water-starworts, which tend to vary with fluctuating water levels, are not dependable for positive identification. In most instances the features of the tiny fruits are the only reliable distinction of the different species (consult manuals for technical keys). Water-starworts are often mistaken for greater duckweed (*Spirodela polyrhiza*, p. 33) or common duckweed (*Lemna minor*, p. 34) due to the roundness of the floating leaves. Close inspection, however, will reveal stems and leaf rosettes among the floating mass of water-starworts and will quickly resolve the confusion between them and the stemless, single, leaf-like bodies of duckweeds.

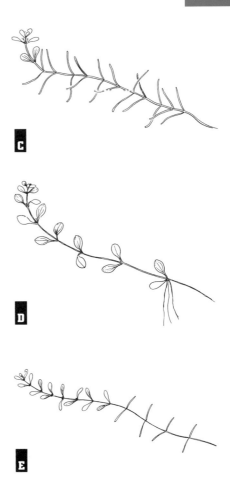

A: *Different-leaf water-starwort* (Callitriche heterophylla, *3.4x magnification*). *As its name suggests, this plant has two leaf shapes—rosettes of rounded, oblong leaves and long, linear leaves.*

B: *Vernal water-starwort* (C. verna).

C: *Different-leaf water-starwort* (C. heterophylla).

D: *Pond water-starwort* (C. stagnalis).

E: *Vernal water-starwort* (C. verna).

FALSE LOOSESTRIFE
WATER-PURSLANE • MARSH SEEDBOX
Ludwigia palustris

Onagraceae (Evening-primrose Family)
INDICATOR STATUS: OBL

GROWTH HABIT: Perennial introduced emergent forb; common semi-aquatic plant that usually grows in dense, thick mats, submerged in water or emerging from it; **stems are succulent, often reddish, lie flat on mud or float in water**, root freely from regularly spaced nodes.

LEAVES: Succulent; **opposite; oval or spatulate, 1–3.5 cm long; leaf stalks (petioles) are nearly as long as blades**.

FLOWERS: Inconspicuous, tucked into leaf axils; no petals, only greenish outer sepals.

FRUITS: Enlarged into four box-like chambers containing seeds when mature; more noticeable than flowers.

HABITAT: In water or on mud along marshy shores of ponds, lakes and slow-moving streams, and in prairie wetlands.

NATURAL HISTORY: False loosestrife appears in mid-spring and blooms from July through September. It is widespread in North America, Eurasia and Africa and is very tolerant of different water levels. It is eaten by muskrats.

SIMILAR SPECIES: The similar foliage and habitats occupied by American brooklime (*Veronica americana*, p. 64) can cause confusion between it and false loosestrife. However, the flowers of these two species are distinctly different from one another. Water-purslane (*Lythrum portula*, also known as *Peplis portula*, OBL) is a very similar introduced species that is becoming well established. It has a 6-sided seed cup (compared to the 4-sided seed cup of false loosestrife), but otherwise these two species are nearly impossible to tell apart.

NOTES: False loosestrife seeds mature in compact, square, partitioned chambers attached along the stem. These 'seedboxes' give it the common name 'marsh seedbox,' especially in the southern U.S.

A: *False loosestrife* (Ludwigia palustris) *in a dense mat.*
B: *False loosestrife in a typical habitat.*
C: *False loosestrife flowers are tucked into the leaf axils.*

Scrophulariaceae (Figwort Family)
INDICATOR STATUS: OBL

GROWTH HABIT: Annual native emergent forb; compact, **amphibious plant with creeping stems and fleshy leaves; stems produce white runners** that root frequently and give rise to new plants.

LEAVES: Fleshy, narrowly elliptical, 5–18 mm long, 2–7 mm wide; leaf stalks (petioles) are 1–8 cm long.

FLOWERS: Borne singly or in groups on stalks that arise from leaf axils; flower stalk, called a 'scape' if single-flowered or a 'peduncle' if several-flowered, is shorter than leaves and tends to droop; flowers are inconspicuous; petals are small, white, pinkish or sometimes blue.

HABITAT: Shallow, still or flowing water in ponds and lakes, muddy shores and stream-banks where water levels drop or fluctuate with tides to expose mud flats, such as along the Columbia River.

NATURAL HISTORY: Mudwort usually blooms as water levels fall. It is found in the interior valleys and floodplains of the Pacific Northwest, especially along the Columbia River, and is widely distributed in North America.

SIMILAR SPECIES: See water-plantains (*Alisma* spp., pp. 71 & 72), pondweeds (*Potamogeton* spp., pp. 41–47) and water chickweed (*Montia fontana*, p. 104).

A: *Mudwort (Limosella aquatica) in a typical habitat—a sandy mud flat associated with a river. Note the small, white flowers.*
B: *Mudwort.*

AMERICAN BROOKLIME • AMERICAN SPEEDWELL
Veronica americana

Scrophulariaceae (Figwort Family)
INDICATOR STATUS: OBL

GROWTH HABIT: Perennial native emergent forb; 5–60 cm tall; **stems are succulent, smooth, hairless, sometimes purplish**, branched near plant base, sometimes erect or ascending; lower stems tend to creep along ground, often root at leafy nodes.

LEAVES: Opposite; lance- to egg-shaped, about 1.5–8 cm wide and 0.5–5 cm long; appear to clasp stem directly (subsessile), but actually have short, distinctly arching leaf stalks (petioles); **leaf edges finely or coarsely saw-toothed (serrate).**

FLOWERS: In groups of 6–30 in long clusters **(racemes)** on stalks (peduncles) from upper leaf axils; **flowers are 4–10 mm across, light bluish violet, with white centers**; light blue petals often have darker lavender stripes.

HABITAT: Moist ground, shallow water of marshes or wet, sandy soil of gravelly streams.

NATURAL HISTORY: American brooklime blooms from May through July. It is a good soil binder, especially at the muddy edges of water where it forms dense colonies.

SIMILAR SPECIES: American brooklime and water speedwell (*V. anagallis-aquatica*, OBL) have slight differences in leaf attachment that help distinguish one from the other. The leaves of American brooklime have short stalks, while water speedwell leaves are attached directly to the stem (sessile) and clasp it. Skullcap speedwell (*V. scutellata*, p. 65) is another similar member of the genus.

A: *American brooklime (Veronica americana) can form extensive mats in semi-aquatic settings.*
B: *Water speedwell (V. anagallis-aquatica) looks very much like American brooklime and occupies similar habitats.*

Scrophulariaceae (Figwort Family)
INDICATOR STATUS: OBL

GROWTH HABIT: Perennial native forb; 10–50 cm tall; erect to ascending, **semi-aquatic or terrestrial plant**; sometimes **reddish or purplish along stems and leaves**; commonly hairless, but a close inspection may reveal tiny, white hairs on stems and leaves; stems can be curved near base, but are generally weakly ascending, often root at lower nodes.

LEAVES: Opposite; attached directly to stem (sessile); long (2–8 cm), **narrow** (2–15 mm wide), **with pointed tips;** sometimes have teeth along edges.

FLOWERS: In racemes that arise on long, **slender stalks (peduncles) from leaf axils; flowers are tiny, bluish,** 6–10 mm across.

HABITAT: Sloughs, slow-moving streams, marshes and ditches.

NATURAL HISTORY: Skullcap speedwell blooms from May through September. It is also found in wetlands at higher elevations.

SIMILAR SPECIES: Many speedwell species look very similar. The long, noticeably narrower, pointed leaves of skullcap speedwell distinguish it from the other related species. In the valleys and floodplains of the Pacific Northwest, purslane speedwell (*V. peregrina*, OBL) is the only wetland species of this genus with white flowers. It is introduced from the eastern United States. American brooklime (*V. americana*, p. 64) and water speedwell (*V. anagallis-aquatica*, p. 64) both have wider, lance-shaped leaves.

Skullcap speedwell (Veronica scutellata).

Apiaceae, also called Umbelliferae (Carrot Family)
INDICATOR STATUS: OBL

Water parsley (Oenanthe sarmentosa) *can be an indicator species for many types of wetlands west of the Cascades.*

GROWTH HABIT: Perennial native emergent forb; up to 1 m tall (though usually shorter); often reclining or scrambling, with stem tips frequently ascending or curled; often roots at stem nodes; stems are fibrous and succulent, but usually weak and soft, developing a dark red tinge late in the season.

LEAVES: Characteristic, **parsley-like, compound arrangement of toothed leaflets with celery-like odor**.

FLOWERS: **In compound umbels**; main umbel is borne on long, ray-like branch; **flowers are white**.

FRUITS: Maroon-colored seeds mature in late summer.

HABITAT: Swamps, meadows, marshes, edges of slow-moving streams and wooded wetlands that are near saturation or are flooded during early spring.

NATURAL HISTORY: Water parsley blooms from mid-June through August. It creates lush vegetation that typically occupies the wettest or lowest areas of wetlands. It is associated with cattail (*Typha latifolia*) and skunk cabbage (*Lysichiton americanum*) in swamps and in transitional areas, such as between shallow and deep-water marshes and shrub swamps.

SIMILAR SPECIES: Water parsley resembles water-hemlock (*Cicuta douglasii*, p. 202), which is very poisonous.

GROWTH HABIT: Perennial native emergent forb; 20–50 cm tall; **semi-aquatic plant with crooked or Y-shaped form**; stems grow from corm-like rhizomes, floating in water if plant is submerged or growing erect out of water if plant is emergent; easily distinguished from grasses by its distinct, round flower clusters.

LEAVES: Ribbon- or strap-like (look like grass or sedge leaves), light to bright green; 20–60 cm long, 4–10 cm wide; **thick and spongy**, with sheaths at bases, larger specimens are Y- or V-shaped towards base; **parallel veins**; **blunt tips**; leaves on submerged plants are limp, float on water surface.

FLOWERS: In spherical, burred flower heads on short stalks that angle away from stout main stalk (peduncle); each plant has 2–5 female (pistillate) heads and 4–8 male (staminate) heads; male heads are smaller and stalkless, and occur above female heads; commonly, male head will be at tip of main stalk.

FRUITS: 2-seeded achenes, in heads that are 1.2–2 cm in diameter.

HABITAT: Generally grows out of water at edges of ponds, lakes and sloughs.

NATURAL HISTORY: Simple-stem bur-reed occurs from late July to early August. It is important to wildlife. Great blue herons and other marsh birds and waterfowl are attracted by bur-reeds. The fruits are eaten by ducks and all parts are heavily grazed by deer and muskrats.

SIMILAR SPECIES: Giant bur-reed (*S. eurycarpum*, OBL) is another species of bur-reed common in the valleys and floodplains of the Pacific Northwest. It is taller (0.5–1.5 m) than simple-stem bur-reed and has wider leaves (6–20 mm) and thicker, stubbier fruits. Also, giant bur-reed stems branch 2–5 times, and the flower cluster (inflorescence) has 5–12 male heads and 2–4 female heads. Giant bur-reed fruits are also larger, and they are in heads that are 2–3.5 cm across. American tapegrass (*Vallisneria americana*, p. 38) also has grass-like leaves that float on the water surface. However, tapegrass leaves are usually narrower

Simple-stem bur-reed (Sparganium emersum) *in flower (left) and with maturing fruit heads (right)*.

(3–10 mm wide), thinner, not as coarse and not as noticeably triangular in cross-section. Also, both bur-reed species have sheaths at the bases of their leaves, while tapegrass leaves do not have sheaths. Cattails (*Typha* spp., p. 68) can be confused with bur-reeds when not in flower, as they have similar leaf shapes, but cattail leaves are generally darker green than bur-reed leaves. When bur-reed is mature (by mid-July), its characteristic, round, stubby, bristly fruit heads distinguish it from any other plant.

CATTAIL
Typha latifolia

Typhaceae (Cattail Family)
INDICATOR STATUS: OBL

GROWTH HABIT: Perennial native emergent forb; **usually 1–3 m tall**; often grows in extensive, pure stands; **stems are stout, cylindrical**, filled with spongy pith.

LEAVES: Strap-like, dark grayish green, about 1–2.5 cm wide; nearly as high as flower cluster, or higher.

FLOWERS: Form compact, thick, cylindrical, **brown spike at top of tall stalk** (peduncle); male flowers form upper portion of spike; female flowers form lower, dense, brown portion of spike; **male and female spikes are separated by 0.5–4 cm gap** (sometimes up to 8 cm); individual female flowers are up to 1.5 cm long.

HABITAT: Shallow, standing or slow-moving water of marshes, lagoons, sloughs, lakes, ponds and pools, roadside ditches and disturbed wet areas.

NATURAL HISTORY: Cattail blooms from late June through August. It can form extensive, pure (monocultural) stands under nutrient-rich (eutrophic) conditions and with disturbed hydrology, but it is usually co-dominant with bulrushes (*Scirpus* spp.). A large number of waterfowl and semi-aquatic mammals are dependent upon these cattail/bulrush communities, which are critical food sources and provide general shelter and nesting habitat. Geese, beavers, muskrats and other wildlife eat the underground stems (rhizomes) of cattails. Cattails provide shelter and nesting cover for marsh wrens and both red-wing and yellow-headed blackbirds. Cattails are also known to have excellent natural water filtering qualities, an important function and value when considering wetland mitigation and conservation.

SIMILAR SPECIES: Narrow-leaf cattail (*T. angustifolia*, OBL) is generally smaller. Its stems are shorter (1–1.5 m tall) and its leaf blades are more narrow (3–10 mm wide). When narrow-leaf cattail is in flower, it is distinguished by a noticeably wider gap between the male and female sections of the flower spike. Cattail can be mistaken for yellow water-flag (*Iris pseudacorus*, p. 69) or bur-reeds (*Sparganium* spp., p. 67) when these species are immature and not in flower.

NOTES: Cattail rhizomes were eaten by Native Americans and the stems and leaves were woven into mats. Cattail can be germinated from seed but the plants remain small for 2–3 years. It is more effectively propagated by root division.

A: *When cattail* (Typha latifolia) *grows in pure (monocultural) stands it often indicates that the natural hydrology has been altered.*

B: *Narrow-leaf cattail* (T. angustifolia) *has a noticeable gap between its male and female flower spikes.*

YELLOW WATER-FLAG • MUD BANANA
Iris pseudacorus

Iridaceae (Iris Family)
INDICATOR STATUS: OBL

GROWTH HABIT: Perennial introduced emergent forb; forms dense clumps that **look much like garden iris**, but are much taller (to 1.5 m); stems are tall, thick and usually branched.

LEAVES: Flat, stiff, **sword-shaped** (1–2 cm wide, sometimes exceeding flower stalks); pale green with whitish hue.

FLOWERS: In few-flowered clusters on stalks to 50 cm tall; **flowers are bright yellow or copper colored, with tubular, greenish neck and black, brown or dark purple stripes at base of petals.**

FRUITS: Capsules, 5–8 cm long; distinctive, resemble hanging bunches of short, green bananas when mature.

HABITAT: Common in silty mud and rocky or sandy areas on shores of lakes and ponds, along streambanks and in ditches.

NATURAL HISTORY: Yellow water-flag blooms from April to August. A native of Europe, it was introduced for ornamental purposes. Yellow water-flag spreads both vegetatively and by seed. It forms dense mats along rivers and small streams and has become a problem along the Lower Columbia River. It also becomes well established in sand along the edges of marshes and lakes. Muskrats eat the rhizomes.

SIMILAR SPECIES: When yellow water-flag is not in flower it can be confused with cattails (*Typha* spp., p. 68) or bur-reeds (*Sparganium* spp., p. 67), which have similar leaves. Yellow water-flag's leaves are usually shorter, stiffer and broader, however, and it is easily identifiable when it has flowers or fruits.

Yellow water-flag (Iris pseudacorus).

WAPATO · BROAD-LEAF ARROWHEAD
Sagittaria latifolia

Alismataceae (Water-plantain Family)
INDICATOR STATUS: OBL

A, B: *Wapato* (Sagittaria latifolia).

GROWTH HABIT: Perennial native emergent forb; **arrowhead-shaped (sagittate) leaves** and flower stalk are tufted from creeping underground stems (rhizomes) with starchy, edible tubers.

LEAVES: Sagittate (arrowhead-shaped), about 25 cm long and 20 cm wide; sometimes submerged, but above water later in summer; submerged leaves sometimes linear, 4–10 mm wide; leaf stalks (petioles) generally long.

FLOWERS: In 2–8 whorls along a specialized stalk (scape) 20–50 cm long; attractive, **white** flowers, approximately 2 cm wide, supported by second, short, ascending, flower stalk (pedicel).

HABITAT: Easily established in mucky substrate at edges of pools, lakes, ponds, lagoons and sloughs.

NATURAL HISTORY: Wapato blooms from middle to late August. It was formerly widespread in the low-elevation wetlands of valleys and floodplains in the Pacific Northwest, but its numbers have been reduced by flood control, the drainage of wetlands for agriculture and the invasion of competing species, especially reed canary-grass (*Phalaris arundinacea*). The most vigorous remaining populations are found along the Columbia River below Bonneville Dam, where tidal fluctuations maintain suitable mudflat habitat. Waterfowl, shorebirds and geese readily eat the underground tubers. Wapato can be grown from seed or the underground tuber can be transplanted, and it is recommended for use in wetland mitigation.

SIMILAR SPECIES: Deltoid balsamroot (*Balsamorhiza deltoidea*, NOL), also called Puget balsamroot, and arrow-leaf balsamroot (*B. sagittata*, NOL) have leaves that are similar to those of wapato, but balsamroots do not usually grow in wetlands and will not tolerate flooding.

NOTES: Wapato's most distinguishing feature is its leaf shape. It was commonly gathered by Native Americans, who originally named it 'wapato.' Wapato is also known as 'delta potato' and is currently cultivated in China and Japan, where the edible tubers are harvested and used in cooking.

Alismataceae (Water-plantain Family)
INDICATOR STATUS: OBL

GROWTH HABIT: Perennial native emergent forb; 0.5–2 m tall; grows from fleshy, vertical, underground stem (corm) that is usually attached to substrate underwater; true stem is lacking, **flower stalk (scape) is prominent.**

LEAVES: A rise from base of plant; **narrowly elliptical or linear**, rarely more than 5 mm wide, often not much wider than leaf stalk (petiole); held stiffly erect out of water; usually grow taller than flowers.

FLOWERS: Crowded together at end of scape **in many-flowered panicle with whorled branches**; lower panicle branches are umbel-like; upper panicle branches are often single-flowered; **each flower has three circular, whitish, pink or purple petals and three green sepals with purplish edges**; there is a circle of leaf-like bracts below each umbel-like flower cluster.

FRUITS: Seeds are flat, round, approximately 2.5 mm across.

HABITAT: Marshy places, mud at water's edge or emerging from water in lakes and ponds.

NATURAL HISTORY: Narrow-leaved water-plantain blooms in mid-June. It may not appear in dry years. A wide variety of animals and birds eat water-plantains.

SIMILAR SPECIES: Narrow-leaved water-plantain is sometimes divided into two varieties —*A. gramineum* var. *gramineum* and *A. gramineum* var. *angustissimum*—but these are commonly regarded as ecological variants of the same species, with their form depending on growing conditions, such as water depth. Lance-leaf water-plantain (*A. lanceolatum*, OBL) is an introduced Eurasian species that is found in North America only in parts of California and Oregon. Its leaves are narrowly lance-shaped, with spatulate tips and wedge-shaped bases, and often exceed the flower cluster. The leaves of American water-plantain (*A. plantago-aquatica*, p. 72) are 3–15 cm wide, broader than those of narrow-leaved water-plantain. Also, they are more ovate-oblong and are usually somewhat submerged or floating on the water surface.

A

B

A: Alisma gramineum *var.* angustissimum *is a variety of narrow-leaved water-plantain that is usually found in standing water and has only one whorl of flowers.*

B: *Lance-leaf water-plantain* (A. lanceolatum).

AMERICAN WATER-PLANTAIN
Alisma plantago-aquatica

Alismataceae (Water-plantain Family)
INDICATOR STATUS: OBL

American water-plantain (Alisma plantago-aquatica).

GROWTH HABIT: Perennial native emergent forb; **flowering heads tower above short, oblong leaves at plant base**; no true stem; leaves and flower stalk (scape) arise from fleshy, vertical, underground stem (corm).

LEAVES: Oblong-oval and somewhat heart-shaped, or sometimes lance-shaped, 3–15 cm wide.

FLOWERS: Clustered at top of tall, slender scape that rises well above leaves; individual **flowers are on many whorled branches** arranged like umbels; **each tiny flower has three petals that are usually white or slightly pink or occasionally deep pink to purple**.

HABITAT: Quiet waters of shallow to deep marshes, swamps and wet meadows; common in soft, muddy substrates along lakeshores and in ditches.

NATURAL HISTORY: American water-plantain blooms from July through August. It often grows alone and frequently in association with cattails (*Typha* spp.) and rushes (*Juncus* spp.). It creates shade and shelter for young fish. Pintails, teals and other waterfowl eat its seeds.

SIMILAR SPECIES: Narrow-leaved water-plantain (*A. gramineum*, p. 71) has narrower leaves (rarely more than 5 mm wide) that are usually held stiffly erect out of the water.

NOTES: American water-plantain can be started from seeds planted in shallow water in the fall.

Polygonaceae (Buckwheat Family)
INDICATOR STATUS: OBL

GROWTH HABIT: Perennial native emergent floating forb; produces roots all along segmented stem; **stems are rounded, succulent and usually prostrate**.

LEAVES: Large (up to 15 cm long), narrowly elliptical to oblong or lance-shaped; usually float on water surface; leaf stalk (petiole) nearly half as long as blade; stipules at leaf base are each 1–2 cm long.

FLOWERS: In 1–2 cone-like heads atop long stalks (peduncles); each head looks like one large flower but is actually a **panicle of many small flowers**; each panicle is 1–3.5 cm long; **flowers are deep pink**, only 4–6 mm across.

HABITAT: Marshes, ponds, lakes, lagoons and swamps.

NATURAL HISTORY: Water smartweed blooms from June through September. It has adapted to a variety of lifestyles; it can grow prostrate or creeping on land or it can be found submerged or floating in water. Water smartweed spreads rapidly if it is exposed to nutrient-rich (eutrophic) conditions, and it can completely cover the surface of open-water habitats and greatly reduce biological diversity. Water smartweed provides shelter and food for a variety of aquatic life forms. The seeds are an important food source for ducks, geese, marsh birds, shorebirds, upland game-birds and small mammals. They are probably most significant to small songbirds.

SIMILAR SPECIES: In the Pacific Northwest there are more than 30 species of plants commonly called smartweed or waterpepper. Water smartweed and water lady's-thumb (*P. coccineum*, OBL), also called water smartweed, are the most truly aquatic. They are difficult to distinguish from one another, except for the structure of their flower panicles. Water lady's-thumb has cylindrical panicles that are larger (at least 4 cm long) than those of water smartweed. There are other superficial differences between these two species that are not easy to establish, and many botanists treat these two as the same species. Waterpepper (*P. hydropiperoides*, p. 74) is also similar.

NOTES: Species of *Polygonum* are generally referred to as 'knotweeds.' They are characterized by swollen nodes on their stems, where the leaves attach, and a distinct papery sheath (called a 'stipule') surrounding these nodes.

A: *Water smartweed* (Polygonum amphibium).
B: *A more leafy variety of water smartweed found in some populations.*

WATERPEPPER
Polygonum hydropiperoides

Polygonaceae (Buckwheat Family)
INDICATOR STATUS: OBL

A

B

C

GROWTH HABIT: Perennial native emergent forb; amphibious; stems are reclining, grow to 1 m long, root freely at nodes.

LEAVES: Many; 5–12 cm long, narrowly to broadly lance-shaped, taper to sharp point; upper leaves often attach directly to main stem; lower leaves usually have short leaf stalk (petiole); **papery stipules on leaf stem near blade have noticeable hairs.**

FLOWERS: Loosely arranged in narrow spikes at end of flower stalk (peduncle); spikes arise singly or in pairs from leaf axils; **flowers are small and greenish, pinkish or white.**

HABITAT: Marshy or muddy edges of ponds, lakes, lagoons and sloughs; tolerant of variable water levels.

NATURAL HISTORY: Waterpepper blooms from July through September. **It is the most common species of the genus** and an important native wetland plant.

SIMILAR SPECIES: In the Pacific Northwest there are more than 30 similar plant species called smartweed or waterpepper. Marshpepper (*P. hydropiper*, OBL) is an introduced weed that has become a pest. Its yellow-green leaves have a biting, peppery taste, and it grows in drier habitats than waterpepper. Lady's-thumb (*P. persicaria*, FACW) and dotted smartweed (*P. punctatum*, OBL) are also very common in Pacific Northwest wetlands. These related species are similar to waterpepper but have shorter, more sparsely distributed hairs on their stipules. Water smartweed (*P. amphibium,* p. 73) is also similar to waterpepper.

NOTES: *Polygonum* species can generally be referred to as 'knotweeds.' They are characterized by swollen nodes on their stems, where the leaves attach, and a distinct papery sheath (called a 'stipule') surrounding these nodes.

A: *Waterpepper* (Polygonum hydropiperoides).
B: *The hairs on the papery sheaths at the stem nodes help identify waterpepper* (P. hydropiperoides).
C: *Lady's-thumb* (P. persicaria). It *has more tightly crowded flowers than waterpepper* (P. hydropiperoides).

Lythraceae (Loosestrife Family)
INDICATOR STATUS: OBL

GROWTH HABIT: Perennial introduced forb; 1–2 m tall; **stems are hairy and somewhat square**, mostly unbranched near base, branched near top, especially when in flower.

LEAVES: Mostly opposite, some are alternate, often occur in threes; **lance-shaped, with pointed tips**; **often have reddish hue**; leaf base more or less heart-shaped or clasping stem.

FLOWERS: In crowded spikes at ends of stalks (peduncles) that arise from upper leaf axils; several spikes typically arise from single axil; **flowers are bright purple-pink with tissue-like appearance**; 5–7 petals, each approximately 7–10 mm long.

HABITAT: Shallow marshes, edges of large ponds and lakes, and riverbanks.

NATURAL HISTORY: Purple loosestrife blooms from July through September. It aggressively invades emergent wetlands and rapidly crowds out native species. It is commonly associated with cattail (*Typha latifolia*), willows (*Salix* spp.) and Douglas' spiraea (*Spiraea douglasii*). Purple loosestrife can be a transitional species between areas of high flooding and the shrubby communities that border them. This attractive but noxious pest has a low wildlife value.

SIMILAR SPECIES: Another loosestrife found in this region is the annual, native species hyssop loosestrife (*L. hyssopifolium*, OBL). It is a pale, waxy-green plant that is usually found in drier habitats. **Hyssop loosestrife flowers are smaller than purple loosestrife flowers**, and usually not as deep red or purple; they are sometimes even white. Both loosestrife species can be confused with fireweed (*Epilobium angustifolium*, FACU+), a non-woody forb, or Douglas'

A

spiraea (p. 181). All four of these species have similarly colored flowers, but Douglas' spiraea is a small-leafed shrub and fireweed has only one spike of flowers. Purple loosestrife usually has several flowering spikes. Also, loosestrife petals are long and tissue-like, while fireweed petals are rounded and smooth.

NOTES: Purple loosestrife was first introduced to the Willamette Valley as an ornamental. It is also invading from adjacent states where it is plentiful. There is an effort underway to eliminate and thwart the further introduction and spread of this plant.

A: *Purple loosestrife* (Lythrum salicaria).
B: *Hyssop loosestrife* (L. hyssopifolium).

B

YELLOW MARSHCRESS • YELLOW BOGCRESS
Rorippa islandica (R. palustris)

Brassicaceae, also called Cruciferae (Mustard Family)
INDICATOR STATUS: OBL

Yellow marshcress (Rorippa islandica) *grows near sloughs, ponds and lakes and on other marshy shores.*

GROWTH HABIT: Annual native emergent forb; 40–100 cm tall; stems are erect, branched from base of plant.

LEAVES: Leaf form can vary between plants and between leaves on same plant; **lower leaves are usually pinnately divided**, up to 17 cm long; **upper leaves are usually smaller, often shallowly lobed or irregularly toothed rather than pinnately divided.**

FLOWERS: Petals are yellow, about 1–2.5 mm wide, spoon-shaped; flower stalks (pedicels) are 3–12 mm long, up-curved.

FRUITS: Broadly egg-shaped (ovate), more than 2 mm wide, 2–12 mm long, about as long as stalk.

HABITAT: Moist depressions, shallow water, ditches and shores of small ponds, pools and sloughs.

NATURAL HISTORY: Yellow marshcress blooms from June through October.

SIMILAR SPECIES: Western yellowcress (*R. curvisiliqua*, p. 107) also has yellow flowers, but its flowers are much smaller (only 1–2 mm wide) than those of yellow marshcress. The fruits of western yellowcress are also narrower (less than 1.5 mm wide) but they are 6–15 mm long. Watercress (*R. nasturtium-aquaticum*, p. 40) has white flowers and usually tolerates greater amounts of flooding.

NOTES: Yellow marshcress is often divided into varieties based on the sizes and shapes of its leaves and fruits and the hairiness of its stems.

Fabaceae, also called Leguminosae (Pea Family)
INDICATOR STATUS: FACW+

A

GROWTH HABIT: Perennial native forb; **sprawling plant with yellow and purple, pea-like flowers**; stems are thin, smooth, 20–50 cm long.

LEAVES: Pinnately compound, divided into 3–6 (usually 5) small leaflets.

FLOWERS: In 3–12-flowered clusters at ends of very long stalks (peduncles); flowers are 2-lipped (like snap-dragon flowers); **top petal (banner) is yellow; side petals (wings) are pinkish to purplish**, fading to nearly white with age; **bottom petals (keels) are yellow, purple-tipped**.

HABITAT: Pond, pool and lake shores and wet prairies; occasionally in wooded wetlands.

NATURAL HISTORY: Seaside trefoil blooms from mid-May through July. It occurs from sea level along the Oregon and Washington coasts to the wet prairies of the interior valleys. Wildlife eat both the seeds and foliage. The seeds are especially important to small mammals, rabbits, pheasants and quails.

SIMILAR SPECIES: Bog trefoil (*L. pinnatus*, p. 115) is much like seaside trefoil but it has yellow and white flowers and is a thicker,

B

stouter plant. Bird's-foot trefoil (*L. corniculatus*, p. 115), an introduced species, sometimes grows in wet places, but it is usually found in more well-drained soils, as well as in disturbed sites. It has bright yellow flowers that are often tinged with red and much smaller, blunter leaflets than the other trefoils.

A, B: *Seaside trefoil* (Lotus formosissimus).

NODDING BEGGARTICKS
STICKTIGHT • NODDING BUR-MARIGOLD
Bidens cernua

Asteraceae, also called Compositae (Sunflower Family)
INDICATOR STATUS: FACW+

A

GROWTH HABIT: Annual native forb; quickly grows to 20–100 cm tall; stems are erect, tall, slender, smooth and somewhat succulent.

LEAVES: Opposite; **linear to lance-shaped,** with **pointed tip** and **rounded base; attached directly to stem (sessile); irregularly spaced teeth** along leaf edges.

FLOWERS: In many, **large (1–5 cm across), yellow, flower-like heads (composite inflorescences)** that are **grouped at ends of long stalks (peduncles) from upper leaf axils;** heads are erect at first, but **soon droop or bend as if heavy;** each head has a central disk of many tiny, tubular flowers, and may have 6–8 broad, petal-like ray flowers radiating around the disk; **disk flowers are brown with yellow stamens; ray flowers are yellow;** there are **two rows of bracts imme-**

diately below each flower-like cluster; outer row is of long, pointed, greenish, **leaf-like bracts that curve downwards away from flower head** (opposite to ray flowers); **inner row is of brown bracts that are pressed upwards or towards flower head.**

FRUITS: Dark brown or black, flattened achenes, **each with 2–4 barbed projections from top; look like large ticks,** hence their common name.

HABITAT: Usually in mud or marshy ground along edges of open water, such as sloughs and ponds; also in slow-moving waters and shrub swamps.

NATURAL HISTORY: Nodding beggarticks blooms from late July to mid-September. The tick-like achenes are easily shed onto passing

B

animals. They get caught in fur (and clothing) and are thereby carried to new locations. Upland game-birds, songbirds, ducks and other water-fowl eat these fruits, although they are not thought to be a favorite or important food plant.

SIMILAR SPECIES: Leafy beggarticks (*B. frondosa*, p. 139) also grows in wetlands, though it is usually found in drier sites than nodding beggarticks and often in disturbed areas. Leafy beggarticks is distinguished by its leaves, which are on long leaf stalks (petioles), and by the inner disks of its flower-like heads, which are orange, rather than yellow as in nodding beggarticks.

NOTES: The genus name *Bidens* means 'two teeth,' and it refers to the pair of hooked bristles on the achenes. (There are sometimes more than two bristles on the achenes of nodding beggarticks.) The species name *cernua* means 'drooping' or 'nodding' and refers to the way the sunflower-like flower heads turn down soon after opening.

C

A: *Nodding beggarticks* (Bidens cernua). *Backwards-bending (reflexed) bracts surround the base of each flower head.*

B: *Nodding beggarticks.*

C: *Nodding beggarticks' tick-like achenes are armed with barbed bristles.*

WESTERN MANNAGRASS
Glyceria occidentalis

Poaceae, also called Gramineae (Grass Family)
INDICATOR STATUS: OBL

GROWTH HABIT: Perennial native emergent grass; up to 1.5 m tall; **stems are weak, lax, 1.5–2 cm thick.**

LEAVES: Narrow (4–13 mm wide), flat, ribbon-like, soft, smooth, often **decumbent, especially when plant grows in water**; base **is flattened**, as though pressed; **sheaths surrounding stem are closed almost to top.**

FLOWERS: Arranged in narrow, **20–50 cm long panicles at top of long flower stalks**; when in flower, **panicles have purplish cast**; many spikelets per panicle; each spikelet is 15–20 cm long, on erect or ascending stalk (pedicel); lower glume is 1.5–3.5 mm long; upper glume is 2.5–5 mm long; 6–13 florets per spikelet; lemma is 3.5–4.5 mm long, with parallel veins.

HABITAT: Shallow water of ponds, pools and lakes, and depressions in wet prairies.

NATURAL HISTORY: Western mannagrass is somewhat uncommon and will not produce flowers if it is entirely submerged. The genus *Glyceria* includes several species that provide food and cover for birds, including many waterfowl, such as gadwalls, mallards and wood ducks, and small seed-eating birds. Muskrats and small mammals eat aquatic species of mannagrass, while deer heavily graze the semi-aquatic species.

SIMILAR SPECIES: Several mannagrasses grow in the Willamette Valley, either as aquatic or semi-aquatic plants. Northern mannagrass (*G. borealis*, OBL) is the most aquatic species of the genus. It grows in deeper water than western mannagrass, to depths of 30 cm or more. Tall mannagrass (*G. elata*, p. 205) and reed mannagrass (*G. grandis*, p. 205) are also similar. Alkali-grasses (*Puccinellia* spp., NOL) can be distinguished from mannagrasses by their leaf sheaths, which are open nearly their entire length; mannagrass sheaths are closed almost to the top.

A: *Western mannagrass* (Glyceria occidentalis).
B: *Western mannagrass inflorescences.*

Poaceae, also called Gramineae (Grass Family)
INDICATOR STATUS: OBL

GROWTH HABIT: Perennial native grass; up to 1.5 m tall; sprawling, grows from short, scaly, creeping underground stems (rhizomes); above-ground stems are 70–150 cm long and 2–3 mm thick, bowed at base (with roots from lower, hairy nodes) but erect towards tip.

LEAVES: Flat, bright yellow-green; 7–25 cm long, 6–20 mm wide; **coarsely rough on upper surface; very sharp or saw-toothed along edges**.

FLOWERS: Arranged **in golden-yellow to purplish panicles that are open, lax and 10–20 cm long**; panicle branches are 2–8 cm long, flexible, consist of single row of spikelets that give branch a 1-sided appearance; spikelets are short-stalked, broadly flattened and bristly, slightly overlap one another; one fertile floret and 1–3 sterile florets per spikelet.

HABITAT: Water up to 1 m deep; common near streams, along muddy borders of ponds and in shallow ponds, marshes, swamps, ditches and intermittent pools of wet prairies.

NATURAL HISTORY: Rice cut-grass is an important native wetland grass. It usually grows in colonies and often forms a dense zone around ponds. Large populations of rice cut-grass can be referred to as 'rice fields.' Muskrats, ducks, marsh wrens and shorebirds eat its seeds, and ducks also pull up and eat the rhizomes.

SIMILAR SPECIES: See mannagrasses (*Glyceria* spp., pp. 80 & 205–6) and reed canary-grass (*Phalaris arundinacea*, p. 143).

NOTES: Rice cut-grass gets its common name from its seeds, which look like rice, and its leaves, which are sandpapery on their upper surface and have sharp, cutting edges. Because rice cut-grass often grows in thick stands and has sharp leaves that can tear clothes and flesh, it can be treacherous to walk through.

A, B: *Rice cut-grass* (Leersia oryzoides).

COLUMBIA SEDGE
Carex aperta

Cyperaceae (Sedge Family)
INDICATOR STATUS: FACW

A

B

GROWTH HABIT: Perennial native grass-like; 20–100 cm tall; upright stems are closely bunched along shallow underground stems (rhizomes).

LEAVES: Upper leaves are flat, 2–6 mm wide; lower leaves are reduced to scales that girdle base of stem.

FLOWERS: In three (sometimes up to six) cylindrical spikes near or at top of stalk (peduncle); upper spikes tend to be male and are usually on long stalks; lower spikes are female, 1–4 cm long, stalkless (sessile), attached directly in leaf axils; **entire flowering cluster (inflorescence) is copper brown to olive green when mature**; one leaf-like bract typically occurs below lowest spike, it sometimes overtops all spikes, but commonly only surpasses first spike.

FRUITS: Perigynia, which contain achenes, are **somewhat inflated, elliptical or nearly round**, 2–3 mm long, **pale copper brown or washed with purple** or with purple or reddish-brown dots; develop **on lower, female spikes**.

HABITAT: Shores of lakes and ponds; occasionally in wet prairies, often in seasonally flooded areas that dry out later in spring.

NATURAL HISTORY: In Oregon, Columbia sedge flowers and fruits in early spring and the fruits are scattered by late spring. It grows chiefly in the floodplains west of the Cascades and along the major rivers north of Salem to the Puget Trough in Washington. It is especially common in the floodplain of the Columbia River. Columbia sedge can also be found in deeper pools in and around Portland especially on Sauvie Island. Waterfowl eat the fruits and use the bunched stems as nesting areas.

SIMILAR SPECIES: See slough sedge (*Carex obnupta*, p. 160).

A: *Columbia sedge* (Carex aperta) *in a typical setting.*
B: *Columbia sedge. The upper spikes of the inflorescence are stalkless. The lowest female spike is on a long stalk.*

GROWTH HABIT: Perennial native emergent grass-like; **recognized by its smooth, shiny leaves and erect, dense, yellowish or light brown fruits**; stems loosely to densely clustered on a network of short, stout, branching underground stems (rhizomes)

LEAVES: Flat, 3–8 mm wide, **unusually smooth** (most sedges have rough leaves); some leaves extend far above flower spikes, while others only slightly exceed them.

FLOWERS: In long spikes widely spaced along stem; individual **spikes are stalkless (sessile) or on short, often inconspicuous stalks**; upper spikes are male, 2–7 cm long; lower spikes are female, 2–7 cm long, 1–2 cm wide at maturity.

FRUITS: In dense spikes of 20–100 perigynia in distinct rows; perigynia are yellowish brown, inflated, lance-shaped, 4–11 mm long and 1.7–3 mm wide; **styles are hardened and usually contorted**.

HABITAT: Bogs, swamps, edges of ponds, lakes and streams; also in backwater areas of old channels and sloughs that are usually flooded all summer; tolerates standing water.

SIMILAR SPECIES: Beaked sedge (*C. utriculata,* sometimes incorrectly called *C. rostrata,* OBL) is easily confused with inflated sedge, but can be distinguished by its fruiting heads, which are reddish brown and oriented at right angles to stem. The fruiting heads of inflated sedge are yellowish brown and angle towards the tip of the stem.

A

B

A, B: *Inflated sedge* (Carex vesicaria *var.* major).

AWL-FRUITED SEDGE • SAW-BEAKED SEDGE
Carex stipata

Cyperaceae (Sedge Family)
INDICATOR STATUS: NOL; provisionally FACW+

A

B

GROWTH HABIT: Perennial native grass-like; **bright green plant with pyramidal flower cluster; stems are broadly triangular, erect in dense clumps, spongy when fresh, easily crushed between fingers when mature.**

LEAVES: Main blades are coarse, flat, 5–11 mm wide; **sheath is flattened and noticeably wrinkled, or puckered** where leaves attach to stem.

FLOWERS: In pyramidal flower cluster (inflorescence) that is 2–10 cm long and 1–3 cm wide; stalk (peduncle) is broadly triangular in cross-section; individual greenish-brown or straw-colored spikes attach directly to peduncle in offset or irregular fashion.

FRUITS: Perigynia, which contain achenes, are lance-shaped, 4–5.2 mm long at maturity.

HABITAT: Shore communities of lakes, ponds, pools, sloughs and streams; also in wetter areas of prairie wetlands and in wetlands from low to moderate elevations.

NATURAL HISTORY: Awl-fruited sedge flowers from May through August.

SIMILAR SPECIES: The sheaths of Cusick's sedge (*C. cusickii*, OBL) have a coppery tinge towards their mouths, and they are not noticeably puckered along the stem. Fox sedge (*C. vulpinoidea*, OBL) is another look-alike, but it only occurs east of the Cascades. Dense sedge (*C. densa*, p. 158) does not have as strongly triangular stems as awl-fruited sedge, and its leaves are generally narrower (3–7 mm wide). However, both awl-fruited sedge and dense sedge have puckered sheaths along the stem (see photo on p. 158).

NOTES: The name 'saw-beaked sedge' comes from the notches near the tips of the tiny achene fruits. 'Awl-fruited sedge' refers to the awl-like shape (narrow throughout, but broader at the base and tapered to a sharp tip) of both the female flower bracts and the perigynia.

A: *Awl-fruited sedge* (Carex stipata) *in a typical habitat along the edge of a pond.*
B: *Awl-fruited sedge.*

Cyperaceae (Sedge Family)
INDICATOR STATUS: OBL

GROWTH HABIT: Perennial native emergent grass-like; small (3–12 cm tall); extensive, delicate, thread-like underground stems (rhizomes) can give rise to very dense mats that **look like green hair**; above-ground shoots are thin **(1 mm thick) and hair- to needle-like.**

LEAVES: Modified into small, thin, **scale-like sheaths that girdle base of flowering stems**; sheaths are pale green or sometimes purplish.

FLOWERS: In single spike at end of stem; spike is 2.5–7 mm long, wider than stem, **appears white when in flower**; 3–15 flowers per spike, each encased in straw-colored scale (with reddish-brown stripes), which gives spike overall scaly appearance.

FRUITS: Dull yellow-white, with 3 stigmas.

HABITAT: Pool and stream edges, standing water of depressions and vernal pools of wet prairies.

NATURAL HISTORY: Needle spike-rush flowers from June through September. It grows in standing water and in mud, and often occurs in wetlands with common duckweed (*Lemna minor*) and bulrushes (*Scirpus* spp.), in the wetter parts of prairies dominated by soft rush (*Juncus effusus*) or in vernal pool communities. Waterfowl eat both the stems and rhizomes, and muskrats eat the roots. Wild turkeys also eat the spikes.

SIMILAR SPECIES: Spike-rushes (*Eleocharis* spp.) are easily identified by the single flower spikes at the ends of the slender, hair-like shoots. At 3–12 cm tall, needle spike-rush is the smallest of the three most common spike-rushes of the Pacific Northwest. Creeping spike-rush (*E. palustris*, p. 162) is large (10–100 cm tall) and stout. Ovate spike-rush (*E. ovata*, p. 86) is intermediate in size (5–50 cm tall).

A: *Needle spike-rush* (Eleocharis acicularis). *Note the small, spiked inflorescences at the tops of the stems.*

B: *Needle spike-rush often looks very hair-like when it grows in water.*

C: *A flower spike of needle spike-rush.*

OVATE SPIKE-RUSH
Eleocharis ovata

Cyperaceae (Sedge Family)
INDICATOR STATUS: OBL

A

GROWTH HABIT: Annual native emergent grass-like; 5–50 cm tall, loosely tufted; **no underground stems (rhizomes)**; aboveground **stems are ribbed, 0.5–2 mm thick.**

LEAVES: Modified as small, thin, purplish sheaths that girdle bases of stems.

FLOWERS: In charcoal-brown, broadly oval to nearly cylindrical end spikes that are 5–13 mm long; usually no less than 40 flowers per spike, more often 50–100; each flower is encased by purplish or brownish scale (with green midstripe) that is usually 1.7–2.5 mm long.

FRUITS: Straw-colored to dark brown, with two or three stigmas.

HABITAT: Wet prairies, shallow water of ponds, muddy swamps and lakeshores; often in standing water.

NATURAL HISTORY: Ovate spike-rush flowers from June to September. A variety of waterfowl eat the clusters of seeds in the spikes.

SIMILAR SPECIES: Creeping spike-rush (*Eleocharis palustris*, p. 162) looks like ovate spike-rush, but it is larger (10-100 cm tall), it is sometimes tufted, and it has underground stems (rhizomes).

B

A: *Ovate spike-rush* (Eleocharis ovata). It *is a small plant, but it is considerably larger than needle spike-rush* (E. acicularis).
B: *A flower spike of ovate spike-rush.*

Cyperaceae (Sedge Family)
INDICATOR STATUS: OBL

GROWTH HABIT: Perennial native grass-like; 70–170 cm tall; **leafy, dark green; stems are stout, slightly triangular (strongly triangular just below flower cluster); leaves spread out from stems in 3-ranked arrangement**.

LEAVES: Coarse, long, 8–15 mm wide, grass-like, with pointed tips and serrated edges; **arise from plant base as well as along stems, usually extend higher than flower heads;** sheaths at leaf bases are tinted reddish purple.

FLOWERS: In many stalkless (sessile) spikelets collected in small clusters at tips of long stalks (peduncles) that together form compound, **umbel-like flower arrangement**; each flower is nestled at base of greenish-black scale that is about 1.5 mm long; a **group of short, leaf-like bracts arise from just below flower cluster**; longest bract is up to 10 cm long.

HABITAT: Mucky areas along streams, in standing water of springs and swamps, and in mud in wooded wetlands.

NATURAL HISTORY: Small-fruited bulrush is widespread in wetlands throughout the Pacific Northwest. Bulrushes are important to wildfowl. Geese and swans eat the shoots, and geese also eat the underground stems (rhizomes) and the leaves. Muskrats eat bulrush seeds.

SIMILAR SPECIES: When small-fruited bulrush is not flowering, its stems and leaves are hard to distinguish from those of big-leaf sedge (*Carex amplifolia*, FACW+), which also has rounded or triangular stems and 3-ranked leaves. The only distinction between them in the non-flowering stage is made by inspecting the sheaths at the bases of the stems. Wool-grass (*Scirpus cyperinus*, OBL), also called flat-sedge bulrush, has densely tufted, narrow leaves and hairy spikelets. Wool-grass also flowers later than small-fruited bulrush.

NOTES: Its common name can be misleading: the fruit (achene) of this species is small, but the plant is very large.

A, B: *Small-fruited bulrush* (Scirpus microcarpus).

C: *Wool-grass* (S. cyperinus) *is common in the Columbia River Gorge and at moderate elevations.*

SOFT-STEM BULRUSH • TULE
Scirpus tabernaemontani (S. validus)

Cyperaceae (Sedge Family)
Indicator Status: OBL

A

B

GROWTH HABIT: Perennial native emergent grass-like; typically 1–3 m tall; **stems are stout, firm, gray-green, round, as much as 3 cm thick at base, thinner near top** (taper to about 2–4 mm thick just below flower clusters).

LEAVES: Highly modified into long sheaths that closely girdle stem bases; blades are short and poorly developed, or lacking entirely.

FLOWERS: In clusters at tips of stems; may seem to arise from side of stem because of stem-like bract that continues above flower cluster and true stem; individual spikelets are 8–15 mm long, stalkless (sessile), grouped a few together at ends of branches of cluster; flowers are born singly in axils of overlapping, reddish-brown scales.

HABITAT: Deep or shallow water, or in muddy or marshy ground around lakes, ponds, streams, prairie wetlands, marshes and wooded wetlands; tolerates alkali conditions.

NATURAL HISTORY: Soft-stem bulrush occurs from June through September. It provides food and cover for fish, muskrats, raccoons and otters. The hard-coated fruits are an important

C

and common food for ducks, shorebirds and marsh birds. Muskrats and geese eat the stems and underground plant parts. Bulrushes also provide nesting cover for waterfowl, marsh wrens and blackbirds.

SIMILAR SPECIES: Hard-stem bulrush (*S. acutus*, OBL) is very difficult to distinguish from soft-stem bulrush. One highly regarded botanist of the Pacific Northwest claims that to distinguish between them you need ripe fruits, 30x magnification, patience and rum. The 'pinch test' may also help.

Soft-stem bulrush stems are easily crushed, while hard-stem bulrush stems are slightly harder to crush. However, this test requires having both species to compare. Also, the flower heads of soft-stem bulrush are reddish brown, while those of hard-stem bulrush are purplish brown. Others claim that the flower scales of hard stem-bulrush are grayish brown with red-brown lines, while the flower scales of soft-stem bulrush are red-brown with brown lines.

D

NOTES: Soft-stem bulrush is the tallest grass-like marsh plant of the Pacific Northwest. The stems of both hard-stem and soft-stem bulrush were woven into mats by Native Americans.

A: *Soft-stem bulrush* (Scirpus tabernaemontani).
B: *Soft-stem bulrush* (S. tabernaemontani) *in a pool created by a natural spring.*
C: *Hard-stem bulrush* (S. acutus).
D: *Cross sections of the stems of hard-stem bulrush* (S. acutus, *left*) *and soft-stem bulrush* (S. tabernaemontani, *right*), *showing the differences in structure.*

POINTED RUSH
Juncus oxymeris

Juncaceae (Rush Family)
INDICATOR STATUS: FACW+

GROWTH HABIT: Perennial native grass-like; 40–90 cm tall; **stems are round**; leaves arise from base and upper stems.

LEAVES: Flattened, with one edge turned towards stem; blades have faint, **segmented cross-markings**.

FLOWERS: In large, loose panicles of 10–70 heads arranged at varying positions on many, open branches; 3–12 flowers per head; **flower parts are greenish yellow to pale brown**, 3–4 mm long; short, leaf-like bract at base of panicle is 3–8 cm long (much shorter than panicle).

FRUITS: Capsules with pointed tips.

HABITAT: Wet meadows and lakeshores, and standing water of pools, depressions and ditches.

NATURAL HISTORY:
Pointed rush is found from May through August. It is common in the lowlands west of the Cascades, especially in the Willamette Valley.

SIMILAR SPECIES:
Pointed rush is fairly distinct, but its flower clusters looks much like those of jointed rush (*J. articulatus*, p. 166) and taper-tipped rush (*J. acuminatus*, p. 164) and its leaves look like those of dagger-leaf rush (*Juncus ensifolius*, p. 92).

A: *Pointed rush* (Juncus oxymeris).
B: *Pointed rush often grows in shallow water.*
C: *Pointed rush leaves are flattened and ribbed, like iris leaves.*

COVILLE'S RUSH
Juncus covillei

Juncaceae (Rush Family)
INDICATOR STATUS: FACW

GROWTH HABIT: Perennial native grass-like; 5–25 cm tall; **stems are slightly flattened**, arise from many underground stems (rhizomes) that run or creep along ground.

LEAVES: 2–4 leaves per stem; **flattened**, narrow, 1–4 mm wide, **grass-like**, parallel to stem, usually as tall as or taller than flower cluster.

FLOWERS: In 1–5 **small, head-like clusters at stem tip**; leaf-like bract extends above flower heads; 3–7 flowers per head; each flower is 3–4 mm long, **has pale to deep brown outer segments (perianth) and greenish central capsule; capsule becomes dark brown at maturity and exceeds perianth by 1 mm**.

HABITAT: Generally in wet prairies and around ponds and lakes; also in coastal habitats.

NATURAL HISTORY: Coville's rush can be found from July through September, though it is somewhat uncommon.

SIMILAR SPECIES: Grass-leaf rush (*J. marginatus*, p. 168) has a much wider leaf (2–6 mm) than Coville's rush, and its distinctive, bulb-like rhizomes distinguish it from any other rush in our region. See also dagger-leaf rush (*J. ensifolius*, p. 92).

A: *This Coville's rush* (Juncus covillei) *was approximately 9 cm tall.*
B: *Coville's rush.*

DAGGER-LEAF RUSH · THREE-STAMEN RUSH
Juncus ensifolius

Juncaceae (Rush Family)
INDICATOR STATUS: FACW

A

GROWTH HABIT: Perennial native grass-like; 20–50 cm tall; elegant wetland plant; **stems are round**; blade-like leaves enclose stems to about one-half stem height.

LEAVES: Flattened, 2–6 mm wide, **iris-like**, with one edge oriented towards stem; blades have conspicuous, **segmented cross-markings**.

FLOWERS: In two to many **rounded heads grouped at tips of stalks**; each head is 5–15 mm across and contains five to many flowers; flowers are 3–3.5 mm long, range from pale **greenish brown to deep purplish brown**; **leafy bract attaches directly below flower heads and is shorter than them.**

FRUITS: Capsules with **abruptly tapered tips**.

HABITAT: Wet meadows and prairies, shallow water of streams, and edges and standing water of ponds, pools, lakes and other depressions at most elevations.

NATURAL HISTORY: Dagger-leaf rush's growing season is from June through August. It often occurs along small streams at higher elevations after logging disturbances.

SIMILAR SPECIES: Bolander's rush (*J. bolanderi*, OBL) has flower heads that are similar to those of dagger-leaf rush, and both species occur in similar habitats, especially around the edges of small lakes, ponds and pools. Their different leaves provide a telling distinction between these two, however. Bolander's rush has round, thin leaves that are unlike the iris-like leaves of dagger-leaf rush. The leaves of both pointed rush (*J. oxymeris*, p. 90) and dagger-leaf rush are wide, flat and iris-like.

B

A: *Dagger-leaf rush* (Juncus ensifolius).
B: *Bolander's rush* (J. bolanderi).

While most people know about the prairies of the Midwest and Great Plains, many are surprised to learn that much of the Willamette Valley was prairie when it was first settled by Euroamericans.

The name 'prairie' refers here to the wet grasslands that developed on clay soils in the Willamette and Umpqua valleys and the southern Puget Trough, south of the glacial outwash plains. Although wetland prairies are best known for tufted hairgrass (*Deschampsia cespitosa*), many other grasses, sedges and herbaceous species are also present.

Tufted hairgrass and red fescue prairie was a major component of a landscape created by a regime of frequent fire from lightning and thousands of years of occupation by Native Americans, who burned much of the valley almost every year to improve hunting and to maintain populations of wild food plants. Prairie occurred on both wet and dry soils, and on bottomlands as well as foothill slopes.

Camases (Camassia *spp.*) *often appear in tufted hairgrass communities in spring.*

Today, only about 800 hectares of wet and dry prairie are thought to remain, just 0.2 percent of an estimated 400,000 hectares in 1850. Place-names like French Prairie, Howell Prairie, Marks Prairie, Gribble Prairie, Red Prairie and Elliott Prairie still exist, but the prairies themselves are forgotten to all but a few. After annual burning ceased in about 1855, woody plants invaded many prairies and gradually converted them to shrublands or forests. As recently as the 1960s, federal and state maps of natural vegetation in the Willamette Valley indicated that it had originally contained coniferous forest.

In 1850 wetland prairies covered about one third of the Willamette Valley, between 120,000 and

Vernal pool communities often contain fragrant popcornflower (Plagiobothrys figuratus), *common downingia* (Downingia elegans) *and coyote thistle* (Eryngium petiolatum).

Creeping spike-rush (Eleocharis palustris) *often forms pure communities in the depressions, ditches and small, shallow canals of wetland prairies.*

160,000 hectares. They have since been reduced to about 400 hectares, 0.3 percent of their historic range. Some wetland prairies were maintained as pastures or hay meadows because seasonal flooding and their clay soils ('white soils') made them practically worthless for other purposes. Wetter sites were ditched or tiled to improve drainage. Flood control and the advent of grass-seed growing changed all this in the 1950s and 1960s, when most of the remaining wetland prairie was converted to agriculture, and the Willamette Valley became the largest grass-seed producer in the nation. In addition, industrial and housing developments destroyed many prairies in urban areas.

In the 1960s and 1970s researchers rediscovered the forgotten prairie as a component of the Willamette Valley landscape. The Willamette Floodplain Research Natural Area (RNA) in Finley National Wildlife Refuge was for several years the only known remnant of wetland prairie. Intensified searches for rare plants in the late 1970s and 1980s located several important prairie remnants west of Eugene, near Corvallis and east of Salem. Currently, the following five sites in these areas have been preserved, and some are being managed with prescribed fire: Willow Creek Preserve, Fern Ridge RNA, Long Tom Area of Critical Environmental Concern, Willamette Floodplain RNA, and Jackson-Frazier Natural Area. Efforts are underway to protect stands at other sites west of Eugene and east of Salem.

Wetland prairies occur on alluvial soils deposited during floods. They are usually separated from streams and rivers by riparian forests and are frequently interspersed with marshes and oak savannas. Because of the impermeable clay soils, the water of winter rains stays at the surface and shallowly floods the prairies until late spring.

Wetland prairies are typically dominated by tufted hairgrass, which covers between 30 and 50 percent of a site, and often have an undulating topography containing low ridges and troughs, or scattered depressions that form vernal pools. Tufted hairgrass typically forms tall pedestals, and the spaces between the pedestals flood with 5–10 cm of standing water for most of the winter. The water dries by

Slough sedge (Carex obnupta) *often forms pure communities in the depressions or wetter sites of wetland prairies.*

late spring, and a showy array of wildflowers fills the gaps. Large ant mounds, built high enough to avoid winter flooding, are also typical on these sites.

Early spring components of the tufted hairgrass community include camases (*Camassia* spp.), Oregon saxifrage (*Saxifraga oregana*), montias (*Montia* spp.) and Willamette Valley bittercress (*Cardamine penduliflora*). By full spring, buttercups (*Ranunculus* spp.), large-leaf avens (*Geum macrophyllum*), northwest cinquefoil (*Potentilla gracilis*), bog

Tufted hairgrass communities typically dry out by mid- to late summer.

trefoil (*Lotus pinnatus*) and Bradshaw's lomatium (*Lomatium bradshawii*) are found, and vernal pools fill with fragrant popcornflower (*Plagiobothrys figuratus*). In early summer, abundant and diverse prairie grasses, such as California oatgrass (*Danthonia californica*), American slough grass (*Beckmannia syzigachne*) and bent-grasses (*Agrostis* spp.), combine with colorful wildflowers, including narrow-leaf blue-eyed-grass (*Sisyrinchium idahoense*), centaury (*Centaurium erythraea*) and the rare timwort (*Cicendia quadrangularis*, also known as *Microcala quadrangularis*). In middle and late summer the composites appear, including asters (*Aster* spp.), Willamette Valley daisy (*Erigeron decumbens*), gumweed (*Grindelia integrifolia*), western marsh cudweed (*Gnaphalium palustre*) and narrow mule's-ears (*Wyethia angustifolia*). Also found late in the season are the owl's-clovers (*Orthocarpus* spp.), and as the vernal pools dry out, common downingia (*Downingia elegans*) and coyote thistle (*Eryngium petiolatum*) replace popcornflower.

The wetter prairie sites contain communities of one-sided sedge/meadow barley (*Carex unilateralis/Hordeum brachyantherum*), slough sedge (*Carex obnupta*) and creeping spike-rush (*Eleocharis palustris*). Vernal pools, which typically retain water longer into the summer, are dominated by annual hairgrass (*Deschampsia danthonioides*), western mannagrass (*Glyceria occidentalis*), common downingia (*Downingia elegans*), coyote thistle (*Eryngium petiolatum*), western marsh cudweed (*Gnaphalium palustre*) and popcornflowers (*Plagiobothrys* spp.).

Tufted hairgrass is a hardy species that can re-populate disturbed sites fairly rapidly. Consequently, a wetland containing a high proportion of tufted hairgrass

Timwort (Cicendia quadrangularis).

Nelson's checker-mallow (Sidalcea nelsoniana).

(50–90 percent), is not necessarily a remnant of wetland prairie. Some of the other indicator species mentioned above, as well as vernal pools and seasonally flooded areas between the pedestals of tufted hairgrass, should also be present.

The wetland prairies of western Oregon's interior valleys, such as the Willamette Valley, are unique. They developed for thousands of years under soil and moisture regimes found nowhere else in the world. Because of the variety of micro-habitats and hydrologic conditions present, they have an unusually high diversity of plant species. Over 100 plant species have been found at Willow Creek Preserve, where a one meter by one meter plot can contain as many as 30 species.

Several species that are adapted specifically to this habitat are endemic to the Willamette Valley—they do not occur naturally anywhere else in the world—and many of them have become rare. Bradshaw's lomatium (*Lomatium bradshawii*) is a federally listed endangered species, Nelson's checker-mallow (*Sidalcea nelsoniana*) is listed as threatened, and white-top aster (*Aster curtus*), Willamette Valley daisy (*Erigeron decumbens*) and peacock larkspur (*Delphinium pavonaceum*) are proposed for listing. Coyote thistle (*Eryngium petiolatum*), Hall's aster (*Aster hallii*), Willamette Valley bittercress (*Cardamine penduliflora*) and meadow sidalcea (*Sidalcea campestris*) are other endemic species that are not as rare. These endemic plants are good indicators of native wetland prairies.

Wetland prairies have already been extirpated from more than 99 percent of their historic range, and it is essential to protect the few remaining sites. They are highly diverse, complex and poorly understood systems, and they provide an extremely productive food base and diverse habitats for many animals. In addition, the endemic plants and animals living in wetland prairies contain unique genes with unknown pharmaceutical or agricultural potential.

The task of protecting wetland prairies will be difficult. Western Oregon's interior valleys already contain two thirds of the state's population, and another million people are expected in the next 20 years, which will only increase pressures from high property values and development interests. Nevertheless, the wetland prairies are an important and threatened part of our natural heritage that needs to be protected for future generations.

Isoetaceae (Quillwort Family)
INDICATOR STATUS: OBL

GROWTH HABIT: Perennial native emergent quillwort; 8–15 cm tall; **looks like clump of short bunchgrass or new shoots of small rush**; stems are reduced to short, bulb-like, underground structures.

LEAVES: Grass-like; green, 3-sided, hollow; 10–20 per plant; erect, collapsing if plant grows in water; **leaf bases swollen, flattened**.

FLOWERS: Quillworts do not have flowers; instead they reproduce by means of **spores** located in spore cases (**sporangia**) **on inner face of swollen leaf bases.**

HABITAT: Vernal pools of wet prairie communities and along edges of sluggish or intermittent creeks.

NATURAL HISTORY: Nuttall's quillwort can be found from mid-April through early June. Most quillworts either grow partially submerged or are amphibious.

SIMILAR SPECIES: There are four other quillwort species in our region: Bolander's quillwort (*I. bolanderi*, OBL), spiny-spore quillwort (*I. echinospora*, OBL), lake quillwort (*I. lacustris*, OBL) and Howell's quillwort (*I. howellii*, OBL). These are mostly distinguished by technical features of the spores and spore cases (see technical manuals for detailed descriptions). Quillworts can be hard to distinguish from the new shoots of grasses and rushes, especially needle spike-rush (*Eleocharis acicularis*, p. 85). However, the bulb-like stem and the spore cases in the leaf bases are distinctive features of quillworts.

NOTES: Quillworts are closely related to clubmosses and horsetails. The genus name *Isoetes* is derived from the Greek *isos*, 'equal,' and *etos*, 'a year,' perhaps because the leaves die back each year.

A

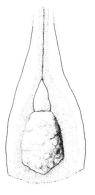

B

A: *Nuttall's quillwort (Isoetes nuttallii). Note the rust-like condition of the roots (rhizophores) along the living root channels.*

B: *The spore cases of Nuttall's quillwort. Quillworts are not flowering plants; instead they reproduce by spores that are produced in the spore cases on the swollen, flattened leaf bases.*

COMMON CAMAS
Camassia quamash

Liliaceae (Lily Family)
INDICATOR STATUS: FACW

A

GROWTH HABIT: Perennial native forb; usually **20–70 cm tall; flowering stalk is taller than leaves**.

LEAVES: Linear, grow from plant base.

FLOWERS: Attached by short stalks (pedicels) at different levels along top of main flower stalk (scape); **flowers are showy, pale to deep blue, 3–5 cm across**; undifferentiated sepals and petals (tepals) are petal-like; bracts at base of each flower are longer than flower stalks (pedicels); albino form produces white flowers.

HABITAT: Wet meadows and prairies.

NATURAL HISTORY: Common camas blooms early, from late April through June, and spreads a beautiful blue over the wet prairies of the Pacific Northwest. In the journal he kept during his expedition with William Clark, Meriwether Lewis described a meadow of camas seen from a distance for the first time as 'resembling a lake of fine clear water.'

SIMILAR SPECIES: Great camas (*C. leichtlinii*, FACW-) is usually taller than common camas, and its bracts are shorter than its flower stalks.

B

In contrast, the bracts of common camas are longer than its short flower stalks. Great camas flowers twist when they wither, while common camas flowers do not usually twist. Also, the seed pods of great camas spread out on long stalks, while the larger seed pods of common camas are pressed close to the main stalk (scape). However, these distinguishing characteristics may apply only to specimens from the Willamette Valley, and not to camas species in the northern valleys of Oregon or in Washington, because these traits vary among specimens of the same species from different geographic locations. The flowers of death-camas (*Zigadenus venenosus*, FAC) are always creamy white, small and crowded together, making it obviously different from common camas and great camas. However, when death-camas is not in flower it can be mistaken for either *Camassia* species, since its leaves and bulbs are very similar to theirs.

NOTES: The onion-like bulbs of common camas were an important food source to several Native American tribes, including the Calapooya in the Willamette Valley and the Nez Perce. However, **deaths** have resulted among people who have eaten death-camas after mistaking it for one of the other camas species.

A: *Common camas* (Camassia quamash). *The bracts just below the flowers are longer than the stubby flower stalks. In contrast, the bracts of great camas are shorter than the flower stalks.*

B: *Common camas* (C. quamash). *The seed pods are pressed close to the main flower stalk (at least in Willamette Valley populations).*

C: *Great camas* (C. leichtlinii). *The seed pods extend well away from the main flower stalk.*

D: *Death-camas* (Zigadenus venenosus) *flowers make it easy to distinguish from common camas and great camas. However, the bulbs of these three species are hard to tell apart and people have died from eating death-camas bulbs that they mistook for camas bulbs.*

HYACINTH BRODIAEA
Triteleia hyacinthina (Brodiaea hyacinthina)

Liliaceae (Lily Family)
INDICATOR STATUS: NOL

GROWTH HABIT: Perennial native forb; **25–70 cm tall**; **onion-like** main stem greatly reduced to underground corm; **white flowers** are on **tall stalk**.

LEAVES: Usually **1–2 at plant base**; **grass-like**, very narrow (1 cm wide), approximately 40 cm long; wither after flowering and can be difficult to find.

FLOWERS: Few to many arranged in umbel at top of slender stalk (scape) that is 25–70 cm tall and surpasses leaves; **flowers are white, with blue or green veins**; bract below umbel is small and papery.

HABITAT: Wet prairies and moist meadows of southern valleys and floodplains.

NATURAL HISTORY: West of the Cascades, hyacinth brodiaea blooms from late July through August. It prefers uplands, but it is often found in wet prairies from the Willamette Valley south to the Rogue River Valley in Oregon. It is also found in sagebrush deserts east of the Cascades.

SIMILAR SPECIES: Elegant brodiaea (*Brodiaea elegans*, FACU), which blooms in late summer, is often found in the prairies of the southern valleys. Like hyacinth brodiaea, it occurs mostly in uplands, but it is often found in wetland prairie associations because it occurs late in the season when these wetlands are drier. Slim-leaf onion (*Allium amplectens*, NOL) is also occasionally found with brodiaeas in wetlands by mid-July, but it is most common in upland habitats. It has very different flowers and its leaves smell of onion when crushed. When camases (*Camassia* spp., p. 98) are not in flower they can be confused with hyacinth brodiaea, but camas leaves are usually much larger, and they do not smell of onions.

A: *Hyacinth brodiaea* (Triteleia hyacinthina).
B: *Elegant brodiaea* (Brodiaea elegans).
C: *Slim-leaf onion* (Allium amplectens).

NARROW-LEAF BLUE-EYED-GRASS
Sisyrinchium idahoense (S. angustifolium)

Liliaceae (Lily Family)
INDICATOR STATUS: FACW-

GROWTH HABIT: Perennial native forb; 15–30 cm tall; **one or more flattened stems per plant**, each with tuft of **iris-like leaves** at base and small but dramatic, **bluish-purple, star-like flowers with yellow centers.**

LEAVES: Grow from bunched bases of each plant, reach approximately midway up flowering stems; flattened and folded like iris leaves.

FLOWERS: One to several; each has six bluish-purple tepals (fused petals and sepals) and small cluster of yellow anthers at center.

HABITAT: Marshes, moist meadows and wet prairies.

NATURAL HISTORY: Narrow-leaf blue-eyed-grass blooms from March through June. It is found throughout the valleys and floodplains of the Pacific Northwest, and is an early spring member of the Willamette Valley wet prairie community.

SIMILAR SPECIES: Hitchcock's purple-eye (*S. hitchcockii*, NOL) is a very similar species found in the same habitats. It is distinguished by its deep reddish-purple flowers. Narrow-leaf blue-eyed-grass tends to have a light blue hue to its purple flowers. Along the coast, golden-eyed-grass (*S. californicum*) grows in bogs and wet inter-dune (deflation plain) areas. It has yellow flowers.

NOTES: Shortly after being picked, the petals of narrow-leaf blue-eyed-grass wilt and shrivel, as though melting.

Narrow-leaf blue-eyed-grass
(Sisyrhinchium idahoense).

CURLY DOCK
Rumex crispus

Polygonaceae (Buckwheat Family)
INDICATOR STATUS: FACW

GROWTH HABIT: Perennial introduced forb; **50–150 cm tall; yellowish green**, often with **reddish or purplish tint over leaves and stems**; single, stout stem is unbranched near base, highly branched above where it supports flowers; papery stipules surround stem above each node and tend to fall off as plant matures.

LEAVES: Long and narrow, with curly edges; upper leaves are smaller than those below.

FLOWERS: Greenish; tepals (undifferentiated petals and sepals) have prominent bump; **flowers become dried, papery, dark reddish to chocolate brown as fruits mature**.

HABITAT: Common in disturbed wetlands.

NATURAL HISTORY: Curly dock is most commonly found from June to September. It was introduced from Europe and is widely established in the valleys and floodplains of the Pacific Northwest. Curly dock is naturalized in most parts of the world and is usually considered to be a bothersome weed. Its seeds are occasionally eaten by waterfowl.

C

SIMILAR SPECIES: Sour dock (*R. acetosella*, FACU+), also called red sorrel, sheep sorrel or sourweed, can be confused with curly dock. However, sour dock is smaller and has small, arrowhead-shaped (sagittate) leaves. Broadleaf dock (*R. obtusifolius*, FAC) and clustered dock (*R. conglomeratus*, FACW-) are also similar, but both these species have much broader leaves than curly dock. Willow dock (*R. salicifolius*, FACW), our only native dock of the species mentioned here, also grows in wet areas and can be confused with curly dock. Willow dock has willow-like leaves that are long and narrow and not crisped around their edges.

NOTES: The stems and fruits of curly dock become brown and appear dead just after flowering.

A: *Curly dock* (Rumex crispus).
B: *Clustered dock* (R. conglomeratus).
C: *Willow dock* (R. salicifolius).
D: *Sour dock* (R. acetosella).

D

WATER CHICKWEED • MINER'S LETTUCE • MONTIA
Montia fontana

Portulacaceae (Purslane Family)
INDICATOR STATUS: OBL

GROWTH HABIT: Annual native emergent forb; **stems are slender, delicate, limp**, 3–20 cm long, branch freely, root at nodes; some plants last for more than one year.

LEAVES: Opposite; egg-shaped (**ovate**) to oblanceolate, narrow, 4–15 mm long, 1–7 mm broad; **attached directly to stems (sessile) or on short leaf stalks (petioles).**

FLOWERS: In 1-sided, drooping clusters (**racemes**) on stalks (**peduncles**) at stem tips; 3–7 flowers per cluster (or sometimes only one); **flowers are delicate**, on slender stalks (pedicels) that are 2–10 mm long; **two sepals, green with white edges, 1 mm long; five petals, white, slightly shorter than sepals.**

FRUITS: Capsules; contain shiny, **jet-black seeds.**

HABITAT: Standing water of meadows and prairies; may also root in streams, often in sandy soils.

NATURAL HISTORY: Water chickweed blooms in April, when willows (*Salix* spp.) are sprouting and early cresses are blooming, and it disappears by late May.

SIMILAR SPECIES: There are two other common valley wetland montias. Narrow-leafed montia (*M. linearis*, NOL) grows 5–20 cm tall, has alternate leaves that are linear, 1.5–5 cm long and commonly 1 mm (but up to 3 mm) wide. The flower clusters (racemes) of narrow-leafed montia are 2–5 cm long and contain 5–12 flowers, each on a stalk that is 5–20 mm long. The leaves of narrow-leafed montia usually grow higher than the flowers. Howell's montia (*M. howellii*, FACW-) is typically 2–6 cm tall and has tiny leaves a mere 5–20 mm long and 0.5–1.5 mm wide.

NOTES: Although water chickweed is not considered a succulent plant, its stems are juicy.

A: *Water chickweed* (Montia fontana).
B: *Narrow-leafed montia* (M. linearis).

WILLAMETTE VALLEY BITTERCRESS
Cardamine penduliflora

Brassicaceae, also called Cruciferae (Mustard Family)
INDICATOR STATUS: OBL

GROWTH HABIT: Perennial native forb; 15–40 cm tall; grows from underground stems (rhizomes) that are tuberous at some nodes; stems are weak, occasionally somewhat prostrate and trailing along ground.

LEAVES: Very few; alternate; narrow; deeply lobed into several paired leaflets; single leaflet at end of leaf points upwards.

FLOWERS: Occur in upper part of plant, tend to droop face downwards; delicate; petals are white, 8–11 mm long.

FRUITS: Long, spreading pods (siliques).

HABITAT: Open wet meadows and prairies, and ash swales; often in standing water.

NATURAL HISTORY: Willamette Valley bittercress appears in early spring, blooms from mid-April through mid-June and soon disappears. Its flowers tend to close up and droop under cloudy skies and during late afternoon and night, and open when exposed to full sun. Willamette Valley bittercress is endemic to the Willamette Valley.

SIMILAR SPECIES: Little western bittercress (*C. oligosperma*, FAC) is an annual plant with smaller flowers (the petals are 2–4 mm long) that do not turn downwards. Watercress (*Rorippa nasturtium-aquaticum*, p. 40) looks remotely like Willamette Valley bittercress, but Willamette Valley bittercress has fewer and more finely divided leaves, and drooping flowers.

A: *Willamette Valley bittercress* (Cardamine penduliflora). *The flowers droop under cloudy skies or in late afternoon.*

B: *The leaflet at the tip of a Willamette Valley bittercress leaf is much larger than the other leaflets.*

RUSH-LEAF COYOTE THISTLE
Eryngium petiolatum

Apiaceae, also called Umbelliferae (Carrot Family)
INDICATOR STATUS: OBL

GROWTH HABIT: Perennial native forb; 15–50 cm tall; **spiny, blue-green,** branched profusely from base; looks like thistly member of sunflower family; stems are widely spreading, **segmented by cross walls, hollow inside;** thick stems bear leaves and flower heads.

LEAVES: Narrow, **lance-shaped,** with **spiny prickles along edges; leaf stalks (petioles) are long and segmented like stem.**

FLOWERS: Many, clustered in stalkless heads that are almost lost in **cluster of green or bluish-tinged bracts;** circle of more **thorny or prickly bracts just below flower head** adds to thistle-like character; **flowers are small, white to greenish, with five petals.**

FRUITS: Mature fruits are green and scaly.

HABITAT: Wet prairies, especially in vernal pools or depressions that are flooded in spring but dry out by late summer.

NATURAL HISTORY: Rush-leaf coyote thistle blooms from July to September. It is found throughout the Willamette Valley, and extends to both sides of the Columbia River up to the east end of the gorge.

SIMILAR SPECIES: Because of its thistle-like character, coyote thistle, especially young plants of it, can be mistaken for needle-leaf navarretia (*Navarretia intertexta*, FACW), which occurs in moist pools to relatively dry places. Needle-leaf navarretia leaves are compound (1–2 times pinnate) and its tiny, blue flowers are tubular.

A: *Rush-leaf coyote thistle* (Eryngium petiolatum) *in its typical habitat with popcorn-flower* (Plagiobothrys *sp.*) *and common downingia* (Downingia elegans).
B: *Rush-leaf coyote thistle* (E. petiolatum).
C: *Needle-leaf navarretia* (Navarretia intertexta).

Brassicaceae, also called Cruciferae (Mustard Family)
INDICATOR STATUS: FACW+

A

GROWTH HABIT: Annual/biennial native emergent forb; **10–40 cm tall**; stems are erect or prostrate, branch diffusely from base or have single main stem that branches mainly near top.

LEAVES: Highly variable in shape; mostly 2–7 cm long, oblong to oblong-lance-shaped; shallowly to deeply **pinnately lobed**; lobes **smooth-edged (entire) to toothed**.

FLOWERS: In small clusters from stem tips and leaf axils; flowers are on short (2–4 mm long), spreading stalks (pedicels); four **yellow petals** are tiny, 1–2 mm long.

FRUITS: Pods (siliques); **6–15 mm long, 1–1.5 mm wide**, usually curved.

HABITAT: Wet prairies, especially in vernal pools, and shores of small ponds and sloughs; also common in ditches.

NATURAL HISTORY: Western yellowcress blooms from May through September. It is also found in higher-elevation wetlands and is native from British Columbia to California and east to Wyoming.

SIMILAR SPECIES: Yellow marshcress (*R. islandica*, p. 76) is a larger plant (30–100 cm tall) with longer leaves (up to 7 cm long).

A: *Western yellowcress* (Rorippa curvisiliqua).
B: *Western yellowcress has curved seed pods (siliques).*

B

OREGON SAXIFRAGE • BOG SAXIFRAGE
Saxifraga oregana

Saxifragaceae (Saxifrage Family)
INDICATOR STATUS: FACW+

GROWTH HABIT: Perennial native forb; 30–120 cm tall; stemless plant, except for **single, long, leafless, sticky flower stalk** (scape) that supports **cluster of white flowers**.

LEAVES: All are at **base of stem**, appear to arise directly from rootstock; linear to lance-shaped or narrowly oval, 7–25 cm long; tapered leaf base; smooth to saw-toothed (serrated) edges; no noticeable leaf stalk (petiole).

FLOWERS: In crowded to open **panicle, 25–125 cm long**, at end of stalk (scape); petals are small, 2–4 mm long, white to greenish, often of unequal lengths; leaf-like bract below panicle; flower stalk is sticky (especially near top) due to **red, gland-tipped hairs**, which can almost be seen without hand lens (see photo).

HABITAT: Wet prairies and moist meadows; also in coastal bogs and occasionally along streambanks.

NATURAL HISTORY: Oregon saxifrage blooms from early spring until June. It is found throughout the Pacific Northwest. It is conspicuous in early spring in association with Willamette Valley bittercress (*Cardamine penduliflora*) and water chickweeds (*Montia* spp.) in tufted hairgrass prairie wetland communities.

SIMILAR SPECIES: Swamp saxifrage (*S. integrifolia*, FACW) differs from Oregon saxifrage by having relatively long (5–40 mm) leaf stalks and a single, head-like cluster of flowers. Rusty-hair saxifrage (*S. ferruginea*, FAC) has oval leaves with coarse, sharp teeth, which distinguish it from Oregon saxifrage.

A: *Oregon saxifrage* (Saxifraga oregana).
B: *Oregon saxifrage's flower stalks have red-tipped glandular hairs* (2x magnification).
C: *Oregon saxifrage has a distinctive rosette of leaves at the base of its stem.*

Ranunculaceae (Buttercup Family)
INDICATOR STATUS: FACW

GROWTH HABIT: Perennial native forb; erect, amphibious, 15–60 cm tall; **one leaf type**; **glossy, yellow flowers** typical of many buttercups; stems are erect, do not root at nodes.

LEAVES: **Simple**, not divided, lobed or dissected as are many other buttercup leaves; **lance-shaped to long and oval**, 4–15 cm long; alternate low on stem, opposite above; usually **smooth-edged (entire)**, but sometimes have slight notches.

FLOWERS: Single or in small groups; each flower has five **bright, glossy, yellow petals**, about **5–10 mm long**, and 25–90 yellow stamens crowded in their centers.

FRUITS: Small, dry, many-seeded achenes.

HABITAT: Wet prairies, moist meadows and ditches, and along banks of ponds and streams.

NATURAL HISTORY: Water-plantain buttercup is an early bloomer in late April through May. Its seeds provide food to upland game birds and waterfowl.

SIMILAR SPECIES: Other wetland buttercups (*Ranunculus* spp., pp. 110–12) and large-leaf avens (*Geum macrophyllum*, p. 113) all have similar yellow flowers (see the field key below).

NOTES: In some locations water-plantain buttercup is called 'dwarf buttercup,' but in the Pacific Northwest it is not at all dwarfed—it grows 15–60 cm tall.

FIELD KEY TO THE COMMON WETLAND BUTTERCUPS
AND LARGE-LEAF AVENS

1a. Leaves are simple, not compound or
deeply lobed . . . . *R. alismifolius*
1b. Leaves are compound or deeply lobed 2
  2a. Basal leaves are pinnately compound 3
    3a. Basal leaves have 9–23 lobed leaflets; end
    leaflet is obviously largest (to 15 cm wide);
    petals are 4–6 mm long . . . . *Geum macrophyllum*
    3b. Basal leaves have 5–7 deeply dissected leaflets;
    petals are 9–18 mm long . . . . *R. orthorhynchus*
  2b. Basal leaves are palmately lobed or compound 4
    4a. Some stems root at their lower nodes . . . . *R. repens*
    4b. Stems do not root at their lower nodes 5
      5a. Stems and leaves are copiously hairy;
      petals are more than 4 mm long . . . *R. occidentalis*
      5b. Stems and leaves are not copiously hairy;
      petals are less than 4 mm long . . . . . . *R. uncinatus*

A, B: *Water-plantain buttercup* (Ranunculus alismifolius).

STRAIGHT-BEAK BUTTERCUP
Ranunculus orthorhynchus

Ranunculaceae (Buttercup Family)
INDICATOR STATUS: FACW-

GROWTH HABIT: Perennial native forb; **20–80 cm tall**; usually **downy or hairy**; **several stems, purplish, slender**, erect or drooping, usually branched.

LEAVES: Three main types; stem-base (basal) leaves are large, divided into **5–7 dissected leaflets** that are 3–8 cm long and deeply cleft (some usually more deeply and finely dissected than others); leaf stalks (petioles) are hairy, up to 25 cm long; **mid-stem leaves are linear**, more shallowly lobed, on shorter leaf stalks; uppermost leaves (**just below flower clusters**) are **stalkless (sessile)**, narrowly 3-parted, with middle lobe usually much longer than other two.

FLOWERS: On stalks (pedicels) that are up to 15 cm long; petals are **yellow, sometimes tinged with orange-red**, 9–18 mm long; **sepals are also yellow, bend back or downwards**.

FRUITS: Achenes; with **straight beak at tip** (most buttercup achenes have curved beaks).

HABITAT: Wet prairies, moist meadows, and along streambanks.

NATURAL HISTORY: Straight-beak buttercup blooms from April to July. It is also found at higher elevations.

D

SIMILAR SPECIES: Hooked buttercup (*R. uncinatus*, FAC-) has similar leaves, but its flowers are smaller (the petals are less than 4 mm long) and its achenes have distinctly hooked beaks. Northwest cinquefoil (*Potentilla gracilis*, p. 114) has a similar yellow flower, but it has palmately divided leaves. See also the field key to buttercups (p. 109).

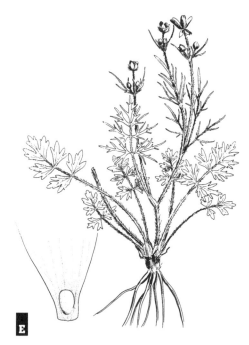

A: *Straight-beak buttercup* (Ranunculus orthorhynchus). *The flowers do not open very wide and the yellow sepals, which bend down towards the stem, fall away after the flower matures.*

B: *Some straight-beak buttercups have simply lobed basal leaves.*

C: *A mature head of straight-beak buttercup fruits (achenes).*

D: *The basal leaves of straight-beak buttercup can be deeply divided and finely dissected.*

E: *Straight-beak buttercup.*

E

WESTERN BUTTERCUP
Ranunculus occidentalis

Ranunculaceae (Buttercup Family)
INDICATOR STATUS: FAC

GROWTH HABIT: Perennial native forb; **15–60 cm tall; one stem**, usually freely branched.

LEAVES: Stem-base (basal) leaves are deeply 3-lobed, with lobes again deeply cleft; leaf stalks (petioles) are thick, 2–10 cm long; upper leaves are attached directly to main stem (sessile) and lobed into three notched segments; **stems and leaves are sparsely to copiously hairy**.

FLOWERS: Typical **shiny yellow buttercup flowers**, 10–25 mm across; **5–8 petals, each more than 4 mm long**.

HABITAT: Moist meadows and wet prairies; often in disturbed areas.

NATURAL HISTORY: Western buttercup blooms in early spring. In coastal habitats it can be found on sun-warmed, rocky slopes. Western buttercup is often found blooming with common camas (*Camassia quamash*) in a beautiful blue and gold display.

SIMILAR SPECIES: Creeping buttercup (*R. repens*, FACW) has creeping, rooting stems (stolons) that are up to 1 m long. It also has slightly larger flowers (10–35 mm across) than western buttercup. Also, see the other wetland buttercups (*Ranunculus* spp., pp. 109–11) and large-leaf avens (*Geum macrophyllum*, p. 113).

NOTES: Because most buttercup flowers look nearly the same, leaf characteristics are the most reliable feature used to identify buttercups. This can get complicated, however, because nearly every species has more than one leaf shape.

A: *Western buttercup* (Ranunculus occidentalis).

B: *Creeping buttercup* (R. repens), *a common garden weed, has leaves that distinguish it from other buttercups.*

C: *Western buttercup* (R. occidentalis).

GROWTH HABIT: Perennial native forb; **20–100 cm tall**; one to several main stems arise from rootstock; **stems are stout, wiry, covered with stiff, bristly, spreading hairs**.

LEAVES: Two leaf types; stem-base (basal) leaves are large (up to 30 cm long), **pinnately compound**, on long leaf stalks (petioles); **9–23 leaflets** of various irregular shapes and sizes; end leaflet is obviously largest (up to 15 cm wide), rounded or somewhat triangular, and often shallowly to deeply lobed around edge; **upper leaves** are variably **3-lobed, like 3-fingered hands**, attached directly to main stems (**sessile**).

FLOWERS: Clustered in upper leaf axils; each flower has **five yellow petals**, each 4–6 mm long; many stamens and pistils in center of flowers give them hairy or feathery appearance typical of rose family flowers.

FRUITS: In bur-like clusters; each rounded fruit (achene) has **hooked style at tip**.

HABITAT: Wetter parts of meadows and wet prairies and shrub swamps.

NATURAL HISTORY: Large-leaf avens blooms in early May.

SIMILAR SPECIES: Due to striking similarities in flowers, various wetland buttercups (*Ranunculus* spp., pp. 109–12) can be confused with large-leaf avens. The long, pinnate leaves at the stem base, and the hooked fruits distinguish large-leaf avens from the buttercups (see the field key on page 109). Large-leaf avens can also be confused with cinquefoils (*Potentilla* spp., p. 114).

A: *Large-leaf avens* (Geum macrophyllum), *showing the stem leaves.*

B: *The basal leaves of large-leaf avens distinguish it from the wetland buttercups.*

NORTHWEST CINQUEFOIL
Potentilla gracilis

Rosaceae (Rose Family)
INDICATOR STATUS: FAC

Northwest cinquefoil (Potentilla gracilis).

FIELD KEY TO THE MOST COMMON WETLAND CINQUEFOILS AND LARGE-LEAF AVENS

1a. Flowers are deep red .... *P. palustris*
1b. Flowers are yellow 2
  2a. Basal leaves are palmately divided .... *P. gracilis*
  2b. Basal leaves are pinnately divided 3
    3a. Leaf undersides are
    silvery-green .... *P. anserina* ssp. *pacifica*
    3b. Leaf undersides are not
    silvery-green .... *Geum macrophyllum*

GROWTH HABIT: Perennial native forb; 40–80 cm tall; graceful, **covered with silky hairs**; each plant has several erect stems that branch extensively in upper sections.

LEAVES: Stem-base (basal) **leaves are all palmately compound**; 7–9 leaflets radiate from central point; leaflets are 2–12 cm long (usually 3–8 cm), deeply indented around edges, sometimes folded; indentation varies considerably from leaf to leaf and from plant to plant; **1–3 stem leaves, stalkless (sessile)**, divided into 3 spreading leaflets that have deep, triangular teeth around edges.

FLOWERS: Clustered in upper branches of main stems, with leafy bracts below them; **flowers are saucer-shaped**, pale to deep yellow, with five petals and many stamens.

HABITAT: Wet prairies, moist meadows and along streams; often in moderately saline soils, moist areas of sagebrush desert, subalpine meadows and grassland habitats.

NATURAL HISTORY: Northwest cinquefoil blooms from May until mid-June. It is common in the wet prairies of the Willamette Valley and ranges from British Columbia to the Dakotas. Northwest cinquefoil is extremely variable and it has been classified into at least seven different varieties.

SIMILAR SPECIES: Pacific silverweed (*P. anserina* ssp. *pacifica*, also known as *P. pacifica*, OBL) is a low-growing species found along the coast and extending up the Columbia River. Its leaves are pinnately compound (feather-like), rather than palmate, and the leaflets have silvery green undersides. Marsh cinquefoil (*P. palustris*, OBL) grows in water, has dark, brownish-red flowers and is found along the coast and at higher elevations. Large-leaf avens (*Geum macrophyllum*, p. 113) has yellow flowers, but it does not have palmately compound leaves like northwest cinquefoil, nor does it have silvery-green leaf undersides like Pacific silverweed.

Fabaceae, also called Leguminosae (Pea Family)
INDICATOR STATUS: FACW

GROWTH HABIT: Perennial native forb; low-growing, with **cream-colored or yellow flowers**; **stems are 20–100 cm long**, can be hairy or hairless.

LEAVES: Compound, divided into **5–9** (usually 7) **oblong leaflets**; leaf stalks (petioles) are often longer than leaves; leaflets are bright green and smooth on upper surface, paler and slightly downy (or sometimes hairless) on undersides.

FLOWERS: In 3–12-flowered, umbel-like clusters at ends of very long stalks (peduncles); flowers are **irregularly shaped, 2-lipped**; **upper lip (banner) and lower lip (keel) are yellow**; **side petals (wings) are white**.

HABITAT: Wet prairies, boggy meadows, swamps, streambanks and ash swales.

NATURAL HISTORY: Bog trefoil blooms from mid-May through June. It occurs west of the Cascades and along the Columbia River Gorge from northwestern Washington to southern California. Bog trefoil grows from sea level to higher elevations in the mountains where it is found mostly along streams. It is a well-established member of the wet prairie and wet woodland communities of the Willamette Valley. Its seeds and foliage provide food for a variety of wildlife species. The seeds are especially important to quail and small mammals.

SIMILAR SPECIES: Bird's-foot trefoil (*L. corniculatus*, FAC) also grows in wet places, but it is an introduced species and is usually found in drier, disturbed sites. Also, bird's-foot trefoil has bright, completely yellow flowers and its leaflets are much smaller and more blunt than the other trefoils. Seaside trefoil (*L. formosissimus*, p. 77) is much like bog trefoil except that seaside trefoil is a thinner plant and usually has some pink, red or purple in its flowers.

A: *Bog trefoil* (Lotus pinnatus).
B: *Bird's-foot trefoil* (L. corniculatus).

LARGE-LEAF LUPINE
Lupinus polyphyllus

Fabaceae, also called Leguminosae (Pea Family)
INDICATOR STATUS: FAC+

Large-leaf lupine (Lupinus polyphyllus).

GROWTH HABIT: Perennial native forb; fairly large lupine with **blue flowers** and one to many hairy stems.

LEAVES: Compound, divided into **5–18 leaflets** that **radiate from a central point**; leaflets are **15–40 cm long**, with smooth upper surfaces and very hairy undersides; **leaves are alternate** along main stem, **far below flowers**.

FLOWERS: Arranged in clusters (**raceme**) along **upper section of main stem**; flowers are **pea-like, bluish purple** (but become brown with age); flower stalks (pedicels) are short.

FRUITS: Pods, **2.5–5 cm long**, covered with dense hairs; open on one side; contain bean-like seeds.

HABITAT: Wet meadows and prairies, swampy places and along streams.

NATURAL HISTORY: Large-leaf lupine blooms in June. Its seeds are a valuable food for game birds.

SIMILAR SPECIES: Riverbank lupine (*L. rivularis*, FAC) is also found in wet prairies and along sandy streambanks. It has white tips on its blue flowers in contrast to the all-blue flowers of large-leaf lupine. There are many other lupine species and most are difficult to distinguish. Large-leaf lupine and riverbank lupine are the most common species in the floodplain wetlands of western Oregon and Washington.

NOTES: The flowers of large-leaf lupine vary greatly from plant to plant, from pale blue to dark violet or purple. Occasionally, all white-flowered plants occur within populations of blue-flowered plants. The species name, *polyphyllus*, literally means 'many-leafed,' but large-leaf lupine is actually 'many-leafleted.'

Apiaceae, also called Umbelliferae (Carrot Family)
INDICATOR STATUS: FACW

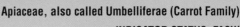

GROWTH HABIT: Perennial native forb; generally 20–65 cm tall; both **leaves and flower stalks (peduncles), grow from base of plant.**

LEAVES: All grow from plant base on leaf stalks (petioles); **divided into fine, highly dissected segments.**

FLOWERS: Yellow; **clustered in compound umbels**; rays of main umbel are usually of **unequal length; secondary umbels are surrounded (subtended) by sets of bracts (involucels); bracts are tiny, leafy, divided into three bractlets**, which are each **further divided into segments**.

HABITAT: Wet, open prairies.

NATURAL HISTORY: Bradshaw's lomatium blooms in early spring, from April to mid-May. It sets seed quickly and the flowers soon fade. Many plants within a population are sterile and never develop seeds. Bradshaw's lomatium is endemic to the wet prairies of western Oregon.

SIMILAR SPECIES: Although the other similar lomatiums of Oregon and Washington usually grow in drier habitats, nine-leaf lomatium (*L. triternatum*, NOL), common lomatium (*L. utriculatum*, NOL), which rarely grows in wetlands, and Cook's lomatium (*L. cookii*), of Jackson and Josephine counties in Oregon, can be confused with Bradshaw's lomatium. Bradshaw's lomatium is best distinguished by the unique structure of the bracts at the bases of the secondary umbels (see photo and illustration).

NOTES: Bradshaw's lomatium is a federally listed endangered species. It was once widespread in the wet prairies of the Willamette and Umpqua valleys, but because of the conversion of its habitat for agricultural and industrial uses, it is now limited to a few sites within Oregon's Marion, Benton, Lane and Douglas counties.

A: *Bradshaw's lomatium* (Lomatium bradshawii).

B: *Bradshaw's lomatium has distinctive, twice 3-parted bracts below its inflorescences.*

C: *An individual Bradshaw's lomatium bract.*

DENSE SPIKE-PRIMROSE
Epilobium densiflorum (Boisduvalia densiflora)

Onagraceae (Evening-primrose Family)
INDICATOR STATUS: FACW-

GROWTH HABIT: Annual native forb; 30–100 cm tall; erect; covered with tiny hairs; main stem single or somewhat branched in upper stem; pretty **pink flowers in axils of upper leaves**.

LEAVES: Crowded, **alternate**; **lower leaves are lance-shaped, 1.5–5 cm long**, attached directly to main stem (**sessile**); **upper leaves are dense, oval to lance-shaped**, with subtle teeth around edges.

FLOWERS: In crowded spikes nestled in axils of upper leaves and attached directly to main stem (sessile); **petals are pale pink to rose or purplish, 3–12 mm long, notched at tips**.

HABITAT: Wet prairies, vernal pools and near ponds; mostly in sites that are wet in spring but dry by late summer.

NATURAL HISTORY: Dense spike-primrose appears in late July and August. It is especially common in the Willamette Valley and is nearly always found in clay soils that have a high capacity to shrink and swell.

SIMILAR SPECIES: Stiff spike-primrose (*E. torreyi*, also known as *Boisduvalia stricta*, FACW) is another native species that also occurs in wet prairies, although it is more common in disturbed sites. It is smaller over-all than dense spike-primrose and its petals are tiny (only 1–3 mm long). Stiff spike-primrose stems are rigid, grayish and highly branched from the base. Also see Watson's willow-herb (*Epilobium ciliatum* ssp. *watsonii*, p. 175).

Dense spike-primrose (Epilobium densiflorum).

Gentianaceae (Gentian Family)
INDICATOR STATUS: FAC-

GROWTH HABIT: Annual introduced forb; 10–50 cm tall; upright, hairless; single main stem or **many stems per plant**; **stems are somewhat thickened at nodes**.

LEAVES: **Stem-base (basal) leaves are in rosette**, obovate or oblong-oblanceolate, 1.5–4 cm long, with 3–5 prominent veins; **upper leaves are opposite**, attached directly to main stem (**sessile**), appear to clasp stem, **tend to fold upwards**.

FLOWERS: Single or in clusters; each flower is 1–1.5 cm long, **tubular**, **salmon to bright rose with yellow throats**; anthers twist together after flowering is past.

HABITAT: Wet meadows and prairies.

NATURAL HISTORY: Centaury blooms from June through August. It was introduced from Europe and has become well established west of the Cascades from northwestern Washington to northern California. It is common in the wet prairies of the Willamette Valley.

SIMILAR SPECIES: Monterey centaury (*C. muhlenbergii*, FACW), a native species, is also found in wet prairies, especially in the Willamette Valley. It is small (3–30 cm tall), slender and single-stemmed. Deptford pink (*Dianthus armeria*, NOL), a European species of the pink family (Caryophyllaceae), grows 20–60 cm tall and has linear leaves and pink to red flowers in congested clusters. It is a fairly common garden annual that often escapes into the wild.

A: *Centaury* (Centaurium erythraea). *The flowers are arranged in umbels.*
B: *Monterey centaury* (C. muhlenbergii).

SMALL-FLOWERED FORGET-ME-NOT
Myosotis laxa

Boraginaceae (Borage Family)
INDICATOR STATUS: OBL

GROWTH HABIT: Perennial native forb; dainty, **10–40 cm tall**; sprawling, sparsely flowered; stems are weak, **prostrate at base** with erect tips that tend to droop.

LEAVES: Oblong to lance-shaped; largest at stem base, 1.5–8 cm long, **3–15 mm wide**; progressively smaller and more linear towards stem tips; attached directly to main stems (**sessile**).

FLOWERS: In long, **coiled or curled clusters (scorpioid cymes)** from stem tips; each flower has **five flat, blue, fused petals attached to orange, ring-like disk**.

HABITAT: Wet prairies, especially in vernal pools, ditches and along edges of pools, ponds and sloughs.

NATURAL HISTORY: Small-flowered forget-me-not blooms from June to September. It grows in clusters at low elevations throughout the Pacific Northwest.

SIMILAR SPECIES: Yellow-and-blue forget-me-not (*M. discolor*, FACW) and true forget-me-not (*M. scorpioides*, FACW) are both introduced species that are also found in shallow water and wet soils throughout the Pacific Northwest. Yellow-and-blue forget-me-not has yellow flowers that turn blue with age. It is more erect, it grows 10–50 cm tall, and it has hairy leaves that are 1–4 cm long and 2–8 mm wide low on the stem and smaller and more scattered above. True forget-me-not is usually a taller plant (20–60 cm) with wider leaves (7–20 mm) than the other two species.

NOTES: The genus name *Myosotis* is from the Greek *mus*, 'a mouse,' and *ous*, 'an ear.' It describes the short, furry leaves of some species (but small-flowered forget-me-not has hairless leaves). The species name of small-flowered forget-me-not, *laxa*, describes its stems, which tend to be limp.

A

A: *Small-flowered forget-me-not* (Myosotis laxa).
B: *True forget-me-not* (M. scorpioides).

B

Boraginaceae (Borage Family)
INDICATOR STATUS: FACW

GROWTH HABIT: Annual native forb; **10–40 cm tall**; usually have one main stem that forks into several branches near top; leaves and stems are sparsely to moderately hairy.

LEAVES: Lower leaves are long, very narrow, in opposite pairs; upper leaves on branching **stems are smaller, alternate;** all are attached directly to stems (sessile).

FLOWERS: In long, coiled, loosely flowered clusters (**racemes**) from stem tips; **flowers are showy, white**, fragrant, **4–10 mm across**, with yellow central disk; sepals are hairier than other plant parts, 3–4 cm long when mature.

FRUITS: Four ovate nutlets, 1.2–1.7 mm long.

HABITAT: Wet prairies, especially vernal pools.

NATURAL HISTORY: Fragrant popcornflower blooms from mid-May through early July. It adapts well to intermittently flooded sites, such as vernal pools, and it is common in the wet prairie communities of the interior valleys of Oregon and Washington, including the Columbia River Gorge. Fragrant popcornflower is also found in the lowlands of British Columbia, but it is not known on the eastern side of the Cascades. Popcornflower plants can either grow singly or in large patches, sometimes covering whole meadows.

SIMILAR SPECIES: Scouler's popcornflower (*P. scouleri*, also known as *P. cognatus*, FACW) looks very much like fragrant popcornflower, but it has smaller flowers. White-flowered varieties of forget-me-not (*Myosotis* spp., p. 120) can resemble popcornflowers, but forget-me-nots have broader, oblong to lance-shaped leaves.

NOTES: Popcornflowers get their name from their white flowers with yellow centers, which resemble popcorn. Fragrant popcornflower becomes sufficiently well established in a few places to form large meadows, which look as if they have been spread with popcorn when the plants are in bloom in spring.

A: *A 'meadow' of fragrant popcornflower* (Plagiobothrys figuratus) *in a wetland prairie of the Willamette Valley.*

B: *Fragrant popcornflower* (P. figuratus).

C: *Scouler's popcornflower* (P. scouleri).

CUT-LEAF WATER-HOREHOUND
AMERICAN BUGLEWEED
Lycopus americanus

Lamiaceae, also called Labiatae (Mint Family)
INDICATOR STATUS: OBL

GROWTH HABIT: Perennial native forb; grows in clusters from slender underground stems (rhizomes); **stems are square**, which is characteristic of mint family.

LEAVES: Opposite along stems; narrow, **triangular**, pointed; **deeply lobed around edges**; taper to short leaf stalk (petiole); **resemble black oak leaves**.

FLOWERS: In dense **whorls in leaf axils**; **flowers are white, tiny, 2–3 mm long**.

HABITAT: Common in wet meadows and along edges of pools, lakes and streams.

NATURAL HISTORY: Cut-leaf water-horehound blooms from late July through August. Its tubers are eaten by muskrats.

SIMILAR SPECIES: Northern bugleweed (*L. uniflorus*, OBL), which grows in similar habitats, has more shallowly lobed leaves that are broader, shorter and much less pointed. Wild mint (*Mentha arvensis*, p. 123) has less deeply lobed leaves and its flowers are lavender, rather than white, and more showy. Also, mint plants give off a strong mint fragrance when crushed, while water-horehounds smell faintly peppery.

NOTES: Cut-leaf water-horehound emits a peppery odor when its leaves are crushed. The genus name *Lycopus* is from the Greek *lykos*, 'wolf,' and *pous*, 'foot.'

A: *Cut-leaf water-horehound* (Lycopus americanus).

B: *Cut-leaf water-horehound* (L. americanus) *flowers are in dense whorls in the upper leaf axils.*

C: *Northern bugleweed* (L. uniflorus).

WILD MINT · FIELD MINT · CANADA MINT
Mentha arvensis

Lamiaceae, also called Labiatae (Mint Family)
INDICATOR STATUS: FACW-

GROWTH HABIT: Perennial native forb; 20–80 cm tall; grows from creeping underground stems (rhizomes); **stems are 4-sided (square)**, thick, with short, whiskery hairs; each leaf axil usually **produces either groups of flowers or another branching stem**.

LEAVES: Opposite; broadly lance-shaped, 2–8 cm long, 4–6 cm wide; **edges are toothed**.

FLOWERS: In crowded whorls in upper leaf axils; flowers are small (4–7 mm long), **2-lipped, pink, rose, violet** or sometimes white.

FRUITS: Brown nutlets, four per flower; enlarge and remain on plant into winter.

HABITAT: Wet meadows and prairies, ditches and open areas in shrub swamps.

NATURAL HISTORY: Wild mint is the **only native mint in the Pacific Northwest**. It is also the most common mint in our region. Its nutlets are an important food source for wildlife.

SIMILAR SPECIES: Peppermint (*M. piperita*, FACU+) and spearmint (*M. spicata*, OBL) have their flowers in spikes at the tips of their stems, which easily distinguishes them from wild mint, water-horehounds (*Lycopus* spp., p. 122) and pennyroyal (*M. pulegium*, p. 124), which all have their flowers nestled in leaf axils. Peppermint and spearmint can be confused with each other, however. Peppermint is 30–80 cm tall, its stems and leaves are often tinged with purple and its leaf stalks (petioles) are 4–15 mm long. Its flower spikes are full and plume-like, and it has a strong, minty odor that can be almost biting or burning on the tongue. In contrast, spearmint is usually taller (30–120 cm), its leaves are bright green, unevenly indented around the edges and stalkless (sessile), and its flower spike is sparse and interrupted. Peppermint and spearmint were intro-

A

B

duced to North America from Europe for herbal and culinary uses. They are now widely cultivated and have become naturalized as weeds.

NOTES: Wild mint tea can be made by steeping a handful of leaves for about five minutes.

A: *Wild mint* (Mentha arvensis).
B: *Peppermint* (M. piperita).
C: *Spearmint* (M. spicata).

C

PENNYROYAL
Mentha pulegium

Lamiaceae, also called Labiatae (Mint Family)
INDICATOR STATUS: OBL

A: *Pennyroyal* (Mentha pulegium).
B: *Pennyroyal in a typical habitat.*

GROWTH HABIT: Perennial introduced forb; **stems are greenish gray**, hairy; **powdery blue to lavender flowers are clustered in leaf axils**.

LEAVES: Occur below each flower cluster, toward tops of stems; short, 1–2.5 cm long, narrow or somewhat oblong, slightly down-turned at tips; stalkless (sessile) or on short leaf stalks (petioles), 1–3 mm long; upper leaves become reduced.

FLOWERS: In dense whorls bunched at intervals from midway along plant to tops of stems; **blue**.

HABITAT: Low marshy ground of wet meadows, prairies, fields, pastures, riverbanks and pond edges; common in disturbed areas, such as impoundments, borrow-pits and ditches.

NATURAL HISTORY: Pennyroyal was introduced from Europe for culinary and medicinal purposes. Because of widespread cultivation, it is now well established and has become a weedy pest in natural habitats in the Pacific Northwest.

SIMILAR SPECIES: See cut-leaf water-horehound (*Lycopus americanus*, p. 122) and wild mint (*M. arvensis*, p. 123).

NOTES: Pennyroyal is widely used in dried flower arrangements and retains its minty fragrance for long periods of time. It also makes a tasty tea and is used as a flea repellent for domestic pets.

Scrophulariaceae (Figwort Family)
INDICATOR STATUS: NOL

GROWTH HABIT: Annual native forb; **10–40 cm tall**; stems are simple near base, slightly branched above, covered with short, usually purple-tinged, **gland-tipped hairs that make them a bit sticky.**

LEAVES: Hairy; attached nearly directly to stems on very short leaf stalks (petioles); **lower leaves are simple**, smooth edged (**entire**); **upper leaves have three notches at their tips** (usually shallow, but may extend to middle of leaf); uppermost leaves hardly separate from bracts of flower cluster.

FLOWERS: In dense, **showy spikes**; flowers are **pink to deep pink, tubular, tightly 2-lipped, only slightly opened**; largely hidden by **showy, petal-like bracts.**

HABITAT: Wetland prairies.

NATURAL HISTORY: Rosy owl's-clover blooms from early June to August. It can also be found in upland habitats and can be an indicator of marginal wetlands and the transitions along upland/wetland boundaries.

SIMILAR SPECIES: Hairy owl's-clover (*Castilleja tenuis*, formerly called *Orthocarpus hispidus*, FACU-) grows with rosy owl's-clover in wetlands, often in drying vernal pools. Hairy owl's-clover is about the same size as rosy owl's-clover, but it is very hairy and has smaller, less showy, white flowers. Parentucellia (*Parentucellia viscosa*, p. 126), which is also in the figwort family, can look much like owl's-clovers, but when parentucellia is in bloom, its yellow flowers easily distinguish it from these two owl's-clovers. Paintbrushes (*Castilleja* spp.) are usually found in upland habitats.

NOTES: Owl's-clovers are unrelated to true clovers, which belong to the pea family.

A: *Rosy owl's-clover* (Orthocarpus bracteosus).
B: *Hairy owl's-clover* (Castilleja tenuis).

PARENTUCELLIA
Parentucellia viscosa

Scrophulariaceae (Figwort Family)
INDICATOR STATUS: FAC-

GROWTH HABIT: Annual introduced forb; **10–70 cm tall**; weedy, erect; **stems are simple**, usually unbranched, **covered with short, gland-tipped hairs that make them slightly sticky.**

LEAVES: Hairy; attached directly to main stems (**sessile**); most are opposite, but upper ones can be alternate or offset (nearly opposite); **lower leaves are simple**, somewhat **narrowly oval**, 1–4 cm long, up to 2 cm wide, with **toothed edges**.

FLOWERS: In leafy spikes at stem tips; flowers are nearly stalkless (**subsessile**), **alternate or offset** along main stem, **yellow, tubular, with two open lips**.

HABITAT: Wet prairies, disturbed wetlands, fields and pastures.

NATURAL HISTORY: Parentucellia blooms from early June through August. It can be an indicator of marginal wetlands or of the transitions along upland/wetland boundaries.

SIMILAR SPECIES: When parentucellia is in bloom, its yellow flowers easily distinguish it from the white flowers of hairy owl's-clover (*Castilleja tenuis*, p. 125) and the pink flowers of rosy owl's-clover (*Orthocarpus bracteosus*, p. 125).

NOTES: Parentucellia is native to the Mediterranean region in Europe.

A: *Parentucellia* (Parentucellia viscosa).
B: *Parentucellia in a typical habitat. Note the characteristic cracks in the damp clay soil, which are due to alternate shrinking and swelling.*

Scrophulariaceae (Figwort Family)
INDICATOR STATUS: OBL

GROWTH HABIT: Annual/perennial native forb; usually **20–80 cm tall**; stems are always erect, but vary from delicate and spindly to robust and leafy.

LEAVES: Opposite; oval, succulent, irregularly toothed along edges, often purplish underneath; lower leaves are on short leaf stalks (petioles); **upper leaves are stalkless (sessile)**, tend to **clasp main stem**.

FLOWERS: On long, slender stalks (pedicels) from upper leaf axils; flowers are deep yellow, 2-lipped, 1–4 cm long; petals are fused into **tube, with red or maroon spots around opening; two brown, hairy ridges run down flared throat of tube**.

HABITAT: Wet prairies and meadows, wet ditches, marshy places, springs, streambanks and pond edges.

NATURAL HISTORY: Seep-spring monkeyflower blooms from March through September, depending on the elevation at which it is growing. It is commonly associated with rushes (*Juncus* spp.) and sedges (*Carex* spp.), and grows at most elevations. Taxonomists classify the many different-sized forms of seep-spring monkeyflower into several varieties. Deer and other mammals may graze on monkeyflowers.

A

SIMILAR SPECIES: Musk monkeyflower (*M. moschatus*, FACW+) can be found in the same places as seep-spring monkeyflower, but there are noticeable differences between them. Musk monkeyflower is smaller, has laxer stems and is covered in gland-tipped hairs. These glands make the plant generally slimy, and give off a musky odor. Toothed monkeyflower (*M. dentatus*, OBL) is chiefly found along the coast in the Pacific Northwest. Although generally smooth, hairless and usually smaller, it looks much like seep-spring monkeyflower.

NOTES: Monkeyflower is so-named because its flower is thought to resemble a grinning monkey face.

A: *Seep-spring monkeyflower* (Mimulus guttatus).
B: *Musk monkeyflower* (M. moschatus).

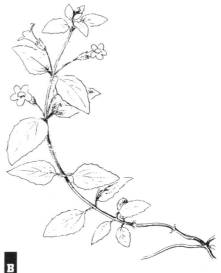

B

BOG ST. JOHN'S-WORT • TINKER'S PENNY
Hypericum anagalloides

Hypericaceae, also called Clusiaceae or Guttiferae (St. John's-wort Family)
INDICATOR STATUS: OBL

Bog St. John's-wort (Hypericum anagalloides).

GROWTH HABIT: Annual/perennial native forb; small, usually **forms mats** by stolons that spread into attractive carpet of ground cover; stems are **slender, prostrate, 3–25 cm long**, root freely at lower nodes; in drier conditions stem tips will sometimes rise above matted vegetation; produces dainty, star-like, yellow to salmon-colored flowers.

LEAVES: Opposite; small, elliptical to rounded, 3–14 mm long, stalkless (sessile), clasp stems; when held up to light and magnified, translucent dots can be seen on leaves.

FLOWERS: 1–5 borne singly or in clusters (**cymes**) either at stem tips or in leaf axils; flowers are yellow to salmon-colored; sepals and petals are rounded.

FRUITS: Tiny, more or less 3 mm long; seeds are yellow-brown, less than 1 mm long.

HABITAT: Many wetland types, including wet prairies, moist meadows and sphagnum bogs.

NATURAL HISTORY: Bog St. John's-wort blooms from late June through early August. It grows at a variety of elevations and in different temperatures and soil conditions. Because it is small, it can go unnoticed under sedges and grasses. Although bog St. John's-wort is not an aquatic plant, it tolerates semi- to permanently flooded conditions. It provides food for ducks and grouse.

SIMILAR SPECIES: Moneywort (*Lysimachia nummularia*, p. 129) is a similar species, but it is much larger.

NOTES: The leaves of bog St. John's-wort give off a **spicy odor** when they are crushed.

Primulaceae (Primrose Family)
INDICATOR STATUS: FACW

■ *Moneywort* (Lysimachia nummularia, *actual size*).

GROWTH HABIT: Perennial introduced forb; often **forms mats** in wet places; stems **creep along ground**, rooting at nodes, **finely dotted with red or black glands**.

LEAVES: Opposite; rounded, 1.5–2.5 cm long and nearly as wide; leaf stalks (petioles) are short.

FLOWERS: Yellow, **1–2.4 cm across**; some are almost as big as leaves.

HABITAT: Wet meadows and prairies, and marshy shores of lakes, ponds and streams.

NATURAL HISTORY: Moneywort blooms from June to August and is well established throughout the Pacific Northwest, including low elevation wetlands on the west side of the Cascade Range. It was introduced from Europe but became naturalized and is now a pest.

SIMILAR SPECIES: Bog St. John's-wort (*Hypericum anagalloides*, p. 128) is another creeping, mat-forming plant with yellow flowers, but it is much smaller than moneywort.

NOTES: Moneywort gets its common name from its smooth, rounded leaves, which resemble coins.

WALL BEDSTRAW
Galium parisiense

Rubiaceae (Madder Family)
INDICATOR STATUS: NOL

Wall *bedstraw* (Galium parisiense).

GROWTH HABIT: Annual introduced forb; weedy, erect or ascending, but tending to become **lax and prostrate** and to **scramble over other vegetation**; stems are slender, square, **10–40 cm long**, unbranched below but irregularly branched from upper leaf nodes, roughened so that they adhere to anything with which they come into contact.

LEAVES: 5–8 in dainty whorls around stems; **linear to linear-oblanceolate**, 4–10 mm long.

FLOWERS: In clustered groups at ends of irregularly branched flower stalks (pedicels); flowers are **tiny, white, with 3–4 fused, petal-like lobes** and 4 stamens arranged between these lobes.

FRUITS: Tiny, **1-seeded nutlets covered with tiny barbed hairs**.

HABITAT: Wet prairies, moist meadows, shrub swamps and especially in wetlands that have been converted to fields and pastures.

NATURAL HISTORY: Wall bedstraw occurs from June through August. It is distributed along the coast from Puget Sound to Oregon, and extends well into the interior valleys to the west slope of the Cascades. The tiny barbed hairs on the fruits easily catch in animal fur or clothing.

SIMILAR SPECIES: There are many other bedstraws found in similar habitats in the Pacific Northwest. The most widespread is common bedstraw (*G. aparine*, FACU), also known as cleavers, which grows in moist, shady places and in drier, upland sites. Also see small bedstraw (*G. trifidum* var. *pacificum*, p. 199).

Campanulaceae (Harebell Family)
INDICATOR STATUS: OBL

GROWTH HABIT: Annual native forb; **10–50 cm tall**; delicate, usually many-flowered; roots are fibrous; stems are thick, single or highly branched.

LEAVES: Linear to lance-shaped, 1–3 cm long, attached directly to stem (**sessile**).

FLOWERS: Single, stalkless (sessile) at ends of stems; flowers are showy, 5–18 mm long, **2-lipped (like snap-dragon flowers)**, with narrow, **flared tube**; **upper lip is blue, with two linear notches at tip**; **lower lip is blue, with white throat** (looks like an eye), surrounded by white border, notched by larger, round and pointed lobes; **anthers are bluish gray, strongly curved, supported by white filaments.**

HABITAT: Vernal pools of wet prairies, edges of ponds and wet meadows.

NATURAL HISTORY: Common downingia blooms from June to August. It is found throughout the lowlands of the Pacific Northwest, from central and eastern Washington and western Oregon to northern California.

SIMILAR SPECIES: Willamette downingia (*D. yina*, OBL), also called Cascade downingia, is smaller, but it is so similar it can be mistaken for common downingia. It is 2.5–7 cm tall and has anthers that are straight or only slightly curved.

A: *Common downingia* (Downingia elegans) *in a typical vernal pool habitat.*
B: *Common downingia.*

TEASEL
Dipsacus fullonum (D. sylvestris)

Dipsacaceae (Teasel Family)
INDICATOR STATUS: FAC

GROWTH HABIT: Biennial introduced forb; 1–2 m tall; stems are coarse, flat-sided, with distinct grooves running down their length; all parts of plant are covered with short, **prickly thorns.**

Teasel *(Dipsacus fullonum). Goldfinches may be responsible for the appearance of this flower head by pecking out the tin,, tubular flowers in an orderly, corn-cob style.*

LEAVES: Opposite, lance-shaped, up to 40 cm long; attached directly to main stem **(sessile);** pair of leaves at stem base often fuse together to form small basin that collects water; mid-veins on undersides of leaves are especially prickly; rosette of basal leaves forms in first year.

FLOWERS: Clustered in dense, **spiked head, 5–7.5 cm long,** surrounded by **5–9 narrow, spine-tipped bracts** that often surpass flower head; **flowers are lavender to pale blue;** flowers in middle of flower head bloom first, followed by those above and below.

FRUITS: 4-sided achene, about 5 mm long.

HABITAT: Disturbed wetlands, pastures, wet roadsides and ditches.

NATURAL HISTORY: Blooming begins with the flowers at the center of the flower spike. After these first flowers fade there are then two bands of blooming flowers on the spike, one above the center and the other below it. Lesser goldfinches reportedly peck out teasel flowers systematically, creating spiral patterns, corn-cob-style, on the flower spike. However, it seems reasonable that this pattern is the result of the flowering pattern described above (see photo). Teasel seeds are known to germinate while still in their seedheads. The water that collects in the cups formed by the fused leaves often houses invertebrates and insects.

SIMILAR SPECIES: Fuller's teasel (*D. sativus*, NOL) is a cultivated species that was introduced from Europe and is now a widespread, opportunistic, noxious weed that displaces more desirable native plants. Thistles (*Cirsium* spp.), which are members of the sunflower family (Asteraceae), vaguely resemble teasels due to their 'thorny' character and bluish flowers, but they do not have opposite leaves.

NOTES: The genus name derives from the Greek *dipsa*, 'thirst,' presumably in reference to the accumulation of water in the fused leaves. The dried flower spike looks like a caged thistle. In Europe the dried flower heads of Fuller's teasel were, and still are, used to raise the nap on wool. Apparently, these heads perform the task so well that man-made tools have not replaced them.

GROWTH HABIT: Perennial native forb; 15–135 cm tall; showy, **lilac-blue flowers**; stems are bushy, somewhat woody at bases, erect, sometimes red-tinted.

LEAVES: Those at stem base are long (up to 25 cm); upper leaves are attached directly to main stems (**sessile**); most have **three distinct veins on upper surface of lower leaves**.

FLOWERS: In flower-like heads that are **single or clustered in groups of up to 20 at ends of long stalks** (peduncles); each head has **20–50 blue or lilac ray flowers** (often mistaken for petals) that are **6–12 mm long**, and **central, yellow disk, 8–15 mm across, of tiny, tubular flowers**.

HABITAT: Open, wet prairies.

NATURAL HISTORY: Willamette Valley daisy blooms from June to early to mid-July. It is endemic to the wet prairies of the Willamette Valley.

SIMILAR SPECIES: See common wetland asters (*Aster* spp., p. 134).

NOTES: Willamette Valley daisy is considered endangered throughout its limited range due to loss of habitat. It is currently listed as an endangered species under Oregon's Endangered Species Act, and is also a candidate for listing under the federal Endangered Species Act.

Willamette Valley daisy (Erigeron decumbens).

A

GROWTH HABIT: Perennial/annual native forbs; typically, one **greenish to purplish** main stem becomes freely branching in upper portion of plants.

LEAVES: Many; most appear to be attached directly to main stems without stalks (**subsessile**); **lance-shaped to oblong**; upper leaves narrower than lower leaves, which often wither early.

FLOWERS: In flower-like heads composed of two types of flowers; outer, strap-like ray flowers (often mistaken for petals) **are blue, pale lavender or violet; central, tiny, tubular disk flowers are usually yellow or dull white.**

HABITAT: Wet prairies, moist meadows, shrub swamps, forested wetlands, and rocky streambanks.

NATURAL HISTORY: Most asters bloom from July through August. Our wetland asters are common at high elevations in the Cascades and on seashores along the coast from Alaska to northern California and eastward through British Columbia and Washington. The flower heads attract bees and butterflies. The central disk of an aster's composite flower head is composed of many tiny tubular flowers. The ray flowers radiate out from the edge of the central disk. They are often mistaken for petals, but they are actually individual flowers with their petals fused into a single strap. These structural complexities and the attendant difficulties in identifying them are why members of the sunflower family are referred to as 'those damned composites.'

SIMILAR SPECIES: Asters look a lot like daisies (*Erigeron* spp., p. 133) but asters tend to have fewer ray flowers in their flower heads. Hall's aster (*A. hallii*, also known as *A. chilensis* ssp. *hallii*, FAC) is most common in the Willamette Valley, but it can be found throughout the interior valleys and floodplains of the Pacific Northwest. Common California aster (*A. chilensis*, FAC) is more widespread in Oregon and Washington, especially along the Columbia River Gorge. These two asters look very similar, but the flower heads of Hall's aster are some-

what smaller on average than those of common California aster. Douglas' aster (*A. subspicatus*, FACW) is common, especially in coastal wetlands. It has similar flower heads but much larger leaves, particularly on the lower parts of the plant, than any of these other asters. White-top aster (*A. curtus*, NOL), which is endemic to the Willamette Valley, has flower heads with only 1–3 white ray flowers that are shorter than the hairs of the pale yellowish disk flowers.

NOTES: White-top aster is a candidate for a 'threatened' or 'endangered' listing on both federal and Oregon endangered species lists, which gives it some protection in the few remaining locations where it occurs.

A: *Hall's aster* (Aster hallii).
B: *White-top aster* (A. curtus).
C: *Common California aster* (A. chilensis).

WESTERN MARSH CUDWEED · LOWLAND CUDWEED
Gnaphalium palustre

Asteraceae, also called Compositae (Sunflower Family)
INDICATOR STATUS: FAC+

GROWTH HABIT: Annual native forb; **3–15 cm tall**; **grayish**; **stems are quite downy** with long, matted, white hairs, and **branch from their bases**.

LEAVES: Even more downy than stems; **alternate**; small, oblong to lance-shaped, mostly 1–3.5 cm long (rarely to 6 cm), 2–10 mm wide.

FLOWERS: In flower-like **heads nestled in highest leaf axils** of main stem; **40–60 tubular flowers per head,** tiny (1.5–2 mm long); **ray flowers are lacking**.

HABITAT: Wet places, especially around vernal pools (it remains long after they have dried up), and undulating and hummocky ground of wet prairies.

NATURAL HISTORY: Western marsh cudweed blooms from June through October throughout the lowlands of the western states. It is tolerant of alkaline habitats.

SIMILAR SPECIES: Marsh cudweed (*G. uliginosum*, also known as *Filaginella uliginosa*, FAC+) looks nearly the same as western marsh cudweed, but it has narrower, more linear leaves. A young pearly everlasting (*Anaphalis margaritacea*, NOL) could be confused with western marsh cudweed, but pearly everlasting is much taller when fully grown and its leaves are not woolly on the upper surface. Tall woolly-heads (*Psilocarphus elatior*, FACW) looks very much like western marsh cudweed, is found in similar habitats, and often grows together with it. However, the leaves of tall woolly-heads are opposite, while western marsh cudweed has alternate leaves.

NOTES: Western marsh cudweed is in the everlasting tribe of the sunflower family.

A: *Western marsh cudweed* (Gnaphalium palustre) *is most common in vernal pools.*
B: *Tall woolly-heads* (Psilocarphus elatior) *is also found in vernal pools and prairie wetlands.*

NARROW MULE'S-EARS
WILD SUNFLOWER • WYETHIA
Wyethia angustifolia

Asteraceae, also called Compositae (Sunflower Family)
INDICATOR STATUS: FACU

GROWTH HABIT: Perennial native forb; **15–80 cm tall**; fragrant; grows from taproot; has **large, showy, yellow sunflowers** and narrow leaves; stems are coarse and leafy.

LEAVES: Stem-base (basal) leaves are **8–50 cm long**, narrow, oblong to lance-shaped, taper gradually at both ends, leaf stalks (petioles) are thick, 1–8 cm wide, 8–50 cm long; **stem leaves are alternate, 5–10 per stem**.

FLOWERS: In flower-like heads; flowers are yellow to almost orange; 10–21 ray flowers (often mistaken for petals), 1.5–3.5 cm long; many disk flowers, tiny, tubular.

HABITAT: Open areas of wet prairies and moist meadows.

NATURAL HISTORY: Narrow mule's-ears is a showy member of the Willamette Valley wet prairie community. It is most common west of the Cascades throughout the Pacific Northwest, but its range extends east of the Cascades to the east end of the Columbia River Gorge. It does especially well in the growing season following a fire.

SIMILAR SPECIES: Arrow-leaf balsamroot (*Balsamorhiza sagittata*, NOL) and deltoid balsamroot (*B. deltoidea*, NOL), also called Puget balsamroot, both have yellow flower heads that are similar to those of narrow mule's-ears. However, the balsamroots are distinguished by their leaves, which have very broad leaf bases. Arrow-leaf balsamroot has arrowhead-shaped (sagittate) leaves and deltoid balsamroot has triangular (deltoid) leaves.

Narrow mule's-ears (Wyethia angustifolia).

Woolly sunflower (Eriophyllum lanatum).

GROWTH HABIT: Perennial native forb; **10–60 cm tall**; showy, **yellow-flowered**; stems are bushy, somewhat woody at bases, can be lax and form thick clumps or sprawling mats.

LEAVES: Variable in size, 1–8 cm long, **covered with velvety, white hairs**; narrow to oblong, **simple** and smooth-edged (**entire**) or shallowly lobed at tips.

FLOWERS: In **sunflower-like heads** typically composed of **5–13 yellow ray flowers** (commonly mistaken for petals) surrounding **central disk of many tiny, yellow, tubular flowers**; ray flowers often **curl downwards** away from disk.

HABITAT: Most common in uplands, especially on well-drained gravel and sand of exposed streambanks; open wet prairies after they have dried out in mid-summer.

NATURAL HISTORY: Woolly sunflower blooms from June through August. In uplands, it often grows on sunny rocky outcrops and screes.

SIMILAR SPECIES: There are many plants with yellow composite flower heads, but woolly sunflower is distinguished by the velvety, white hairs that cover most of the plant. Sneezeweed (*Helenium autumnale*, FACW) also grows on streambanks and other low, moist ground, but it is a much taller plant (15–120 cm) and its 10–20 ray flowers bend back from the disk. Narrow mule's-ears (*Wyethia angustifolia*, p. 137) has much larger leaves and flower heads. Other yellow sunflowers and dandelions include cat's ears (*Hypochaeris radicata*, FACU), hawkweeds (*Hieracium* spp., NOL), smooth hawksbeard (*Crepis capillaris*, FACU), field sow-thistle (*Sonchus arvensis*, FAC-), annual sow-thistle (*Sonchus oleraceus*, UPL), nipplewort (*Lapsana communis*, NOL), common blue lettuce (*Lactuca tatarica* ssp. *pulchella*, also known as *L. pulchella*, FAC), dandelion (*Taraxacum officinale*, NOL), and salsify (*Tragopogon dubius*, NOL). Consult technical manuals for more detailed descriptions.

Asteraceae, also called Compositae (Sunflower Family)
INDICATOR STATUS: FACW+

GROWTH HABIT: Annual introduced forb; 50–150 cm tall; fast-growing; stems are ascending, 4-sided, smooth and somewhat succulent.

LEAVES: Opposite; **pinnately compound, divided into 3–5 rather thin leaflets**; leaflets are lance-shaped, 2–8 cm long, smooth-edged (entire) or with regularly spaced serrations; leaf stalks (petioles) are 1–6 cm long.

FLOWERS: In orange, roundish, flower-like heads (up to 1 cm wide) at tops of long stalks (peduncles) from leaf axils of upper main stem; heads composed of small, tubular disk flowers and (sometimes) strap-like ray flowers; bracts are in row below flower head, leaf-like, long (typically surpass disk flowers).

FRUITS: Dark brown or black achenes; distinctively **flattened**, with **two barbed projections at top**; resembles a tick (see drawing).

HABITAT: Wet places associated with marshy shores and prairies; sometimes in relatively dry disturbed areas.

NATURAL HISTORY: Leafy beggarticks blooms from late June through October. Upland gamebirds, songbirds, and waterfowl eat the achenes. The barbs on the achenes catch in animal fur and clothing.

SIMILAR SPECIES: Lobed beggarticks (*B. tripartita*, FACW), also called three-lobed bidens, has leaves that are simple but so deeply lobed that they almost look compound. Nodding beggarticks (*B. cernua*, p. 78) is usually found in wetter sites than leafy beggarticks. The leaves and flower heads of these two species also differ. The leaves of leafy beggarticks are on long stalks and are compound, while the leaves of nodding beggarticks are stalkless and are not compound. Also, the disk flowers of leafy beggarticks tend to be orange, while the disk flowers of nodding beggarticks are more yellow.

NOTES: Presumably, the tick-like appearance of the achenes and their tendency to catch on passing animals are why members of this genus are referred to as beggarticks. *Bidens* means 'two teeth,' and refers to the two barbs on the achenes; *frondosa* means 'leafy.'

A: *Leafy beggarticks* (Bidens frondosa).

B: *Lobed bidens* (B. tripartita) *has more deeply lobed leaves than leafy beggarticks.*

C: *Leafy beggarticks* (B. frondosa). *The involucral bracts surround the flower head.*

D: *A tick-like achene of leafy beggarticks* (B. frondosa).

GUMWEED
Grindelia integrifolia

GROWTH HABIT: Perennial native forb; **15–80 cm tall**; multiple stems are reddish, tough, wiry; **upper stems branch out to support flower heads; stems and leaves are covered with aromatic, resinous or gummy substance**.

LEAVES: Lower leaves are up to 40 cm long (but typically no longer than 20 cm) **and 4 cm wide**, oblong and lance-shaped, arise directly from plant base on long leaf stalks (petioles); smooth-edged (entire) or sometimes toothed; gummier **upper leaves are much smaller**, lance-shaped, stalkless (**sessile**) and clasp main stem.

FLOWERS: In single or grouped **flower-like heads** that are **2.5–4 cm across; ray flowers are yellow**, radiate around **central, yellow disk** that is commonly **1–3 cm across**; flower heads sit on wide, rounded, **green bases with whorls of small, overlapping bracts** that curl away from flower head; the resin is especially thick in base of flower head (often the stickiest part of the plant).

HABITAT: Wet prairies, ditches and disturbed wetlands.

NATURAL HISTORY: Gumweed blooms from July to August. It is well adapted to clayey soils. The aromatic resin is most abundant late in the season, and it often scents the summer air.

SIMILAR SPECIES: Mountain tarweed (*Madia glomerata*, FACU-) produces an even stronger, aromatic, sticky resin and has small yellow flowers. It is normally found in uplands and has much smaller leaves and flowers than gumweed. Mountain tarweed is also extremely hairy, while gumweed appears smooth and shiny. Gumweed can be mistaken for other yellow-flowered composites, but its stickiness and the hooked bracts below its flower heads will distinguish it from most other species of the sunflower family.

A: *Gumweed* (Grindelia integrifolia).
B: *A more leafy form of gumweed.*

Poaceae, also called Gramineae (Grass Family)
INDICATOR STATUS: FAC

GROWTH HABIT: Perennial introduced grass; 0.5–1 m tall; greenish to grayish plant, appears **lax and velvety**; grows in low, dense **tufts**; stems (culms) are covered with soft hairs along entire length, sometimes prostrate and creeping.

LEAVES: Lax, flat, 5–10 cm long, 4–9 mm wide; blades covered with **velvety hairs**.

FLOWERS: In rather **soft, congested panicle**, 7–15 cm long, 4–9 cm wide; panicle is **purplish gray in spring**, becomes bleached or **straw colored by mid-summer**.

HABITAT: Common in wet meadows, prairies, fields, pastures and other disturbed habitats.

NATURAL HISTORY: Velvet grass flowers in early spring and forms seed in July. Although the U.S. Fish and Wildlife Service lists velvet grass as a native to our region, it is actually a weedy European species that was probably introduced as a meadow grass. It spreads by underground stems (rhizomes) and has become a nuisance, but its seeds provide food for lazuli buntings and lesser goldfinches.

SIMILAR SPECIES: No other wetland grass species in our area has quite the character of the abundant soft hairs on the stems of velvet grass. Creeping velvet grass (*H. mollis*, FACU) is 30–80 cm tall, half the size of velvet grass, and its stems are usually bowed at their bases (decumbent) and are only hairy at the nodes. The leaves of creeping velvet grass are 3–15 cm long and 4–9 mm wide, nearly the same dimensions as velvet grass. Also see witchgrasses (*Panicum* spp., p. 155), Kentucky bluegrass (*Poa pratensis*, p. 152), colonial bentgrass (*Agrostis capillaris*, p. 156) and bromes (*Bromus* spp., consult technical manuals).

A: *Velvet grass* (Holcus lanatus) *seed heads develop in spring.*

B: *Velvet grass produces seeds fairly early in the growing season, and by mid-summer its seed heads become straw-colored.*

LARGE BARNYARD GRASS
WATERGRASS • HOG-GRASS
Echinochloa crus-galli

Poaceae, also called Gramineae (Grass Family)
INDICATOR STATUS: FACW

GROWTH HABIT: Annual introduced grass; 60–160 cm tall; stems (culms) can be spongy, especially at bases and between nodes, usually **flattened** at bases with erect tips (**decumbent**), looking as though they have been trampled.

LEAVES: Flat, **1–30 cm long, 6–20 mm wide**, most arise from stem base; **no ligules**.

FLOWERS: In panicles that are 6–10 cm long; spikelets have **long awns** and are quite **hairy**, are clustered along **one side of panicle branches**.

HABITAT: Wet meadows and prairies, ditches, sloughs, disturbed ponds, lakeshores and other disturbed areas.

NATURAL HISTORY: Large barnyard grass forms seed from June through October. It is a European grass that was originally introduced for livestock forage and has since become a weedy species usually found on irrigated sites. The large, smooth seeds are choice waterfowl food, and it was once common practice in waterfowl management to seed large barnyard grass onto moist mud flats, together with water smartweed (*Polygonum amphibium*). Large barnyard grass is now well established in wet areas and has become a management problem because it overtakes and crowds out native plants.

SIMILAR SPECIES: Large barnyard grass can be mistaken for western mannagrass (*Glyceria occidentalis*, p. 80) which also has stems with flattened bases (decumbent). Large barnyard grass is normally quite easy to distinguish from other grasses because of its spongy, sometimes flattened leaves and the hairy aspect of its spikelets. Also, it is the only grass that does not have ligules (the membranous flap on the upper surface of a leaf where the blade joins the stem sheath).

NOTES: The genus name *Echinochloa* comes from the Greek *echinos*, 'hedgehog,' and *chloa*, 'grass.'

A: *Large barnyard grass* (Echinochloa crus-galli).
B: *Basal leaves of large barnyard grass.*

Poaceae, also called Gramineae (Grass Family)
INDICATOR STATUS: FACW+

GROWTH HABIT: Perennial introduced grass; 0.5–1.5 m tall; light-green to straw-colored, with long, wide leaves and swollen stem nodes; stems (culms) are erect, up to 1 cm thick, become reddish near top.

LEAVES: Green with **sooty-gray hue**; **flat**; 10–30 cm long, 7–20 mm wide; tend to **spread out from stem at right angles**; **auricles surround stem** at top of sheath.

FLOWERS: In **narrow, flattened, spike-like panicles** at stem tips; spikelets are densely packed or spaced (near bottom); panicles are **pale green or tinged with purple** in spring, fade to straw color in late summer.

HABITAT: Muddy shores and shallow waters of ponds, lakes, sloughs, ditches and streams.

NATURAL HISTORY: Reed canary-grass is common in disturbed sites, such as roadside ditches, rights-of-way and around impounded water in areas where water levels fluctuate. Many consider it a pest that out-competes other species and forms monocultures. Although the U.S. Fish and Wildlife's *National List of Plant Species* designates it as native to Region 9, it is probably not native west of the Cascades. The form found west of the Cascades (and now throughout most of North America) is thought to be an aggressive cultivar introduced for wetland forage and erosion control. It is very difficult to eradicate since it propagates both by seed and by vegetative runners that are only encouraged when attempts are made to uproot the main plant. Reed canary-grass has a high tolerance to salt and nitrates. It provides food, cover and nesting habitat for waterfowl, marsh birds and small mammals.

SIMILAR SPECIES: The **clasping auricles** of reed canary-grass help distinguish it from similar species. Harding grass (*P. aquatica*, FACU+) has cylindrical panicles that are smaller, denser and more compact than those of reed canary-grass. Common reedgrass (*Phragmites australis*, FACW+), which is widespread in the eastern states but not as common in the Northwest, looks much like reed canary-grass but is much larger, generally 2–3 m tall. See also western mannagrass (*Glyceria occidentalis,* p. 80), and blue-joint (*Calamagrostis canadensis,* p. 151).

NOTES: Its species name, *arundinacea*, means 'reed-like,' and appropriately describes this large grass. Reed canary-grass is also thought to look like bamboo because its long, wide leaves spread at right angles from the swollen stem nodes, and because it is light green to straw-colored.

A: *Reed canary-grass* (Phalaris arundinacea).
B: *Harding grass* (P. aquatica).

MEADOW FOXTAIL
Alopecurus pratensis

Poaceae, also called Gramineae (Grass Family)
INDICATOR STATUS: FACW

GROWTH HABIT: Perennial introduced grass; 48–77 cm tall, or sometimes trailing on surface of shallow water; plants grow individually or in **dense clusters**; **stems are smooth, hairless, slender**, tend to root at lower nodes; stems of terrestrial plants are often reclining or erect with slight bend at lower nodes; stems of submerged plants float on water.

LEAVES: Up to 10 cm long, 3–10 mm wide, slightly **rough** to touch.

FLOWERS: In dense, narrow, erect, **cylindrical panicles, 3–10 cm long, at stem tips**; panicles appear bristly due to **straight awns (more than 4 mm long)** at tips of lemmas.

HABITAT: Muddy borders of ponds, river bottoms, sloughs, wet meadows, marshes and swampy places.

NATURAL HISTORY: Meadow foxtail was introduced from Europe, and it is now common either in areas of seasonally standing water or on drier ground. In early spring, just before the flowers start producing pollen, the panicles are a lovely shade of bluish lavender. Once the flowers start producing pollen (anthesis) the panicles turn a bright orange to burnt copper color and appear shaggy. These phases often occur at slightly different times in plants that are near one another, which creates a rather dramatic effect.

SIMILAR SPECIES: Water foxtail (*A. geniculatus*, FACW+) and short-awned foxtail (*A. aequalis*, OBL) are two other foxtails that are found in wetlands in our region. Water foxtail grows 14–55 cm tall and is more aquatic. Its awns are bent and are also less than 4 mm long. Short-awned foxtail is the smallest of the three (9–47 cm tall) and its stems bend noticeably at their lower nodes and tend to recline. It has straight awns that are less than 4 mm long. Timothy (*Phleum pratense*, FAC-) can also be confused with meadow foxtail, but its stems are much more bunched at their bases and its seeds also differ (see drawings). Although timothy can grow in moist places, it is not as likely to be found in areas that are as wet as those meadow foxtail tolcrates. In early spring, before the panicles develop, reed canary-grass (*Phalaris arundinacea,* p. 143) can be mistaken for meadow foxtail, but meadow foxtail can be distinguished by its leaves, which feel rough when the blade is pulled through the fingers from the base to the tip. When mature, reed canary-grass is much stouter, taller and leafier than meadow foxtail, and it has a compressed or open panicle.

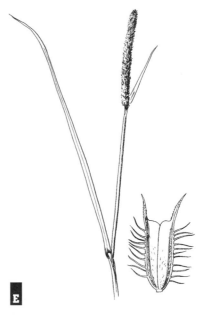

A: *Meadow foxtail* (Alopecurus pratensis) *flower spikes are dramatically different colors at different stages in development. The bluish spike (left) has not yet produced pollen (pre-anthesis) while the copper-colored spike (right) has finished producing pollen (post-anthesis).*

B: *Meadow foxtail* (A. pratensis).

C: *A floret of meadow foxtail* (A. pratensis).

D: *Water foxtail* (A. geniculatus).

E: *Timothy* (Phleum pratense).

MEADOW BARLEY
Hordeum brachyantherum

Poaceae, also called Gramineae (Grass Family)
INDICATOR STATUS: FACW-

GROWTH HABIT: Perennial native grass; 40–80 cm tall; grows in **discrete tufts**; stems (culms) are erect, **smooth, nearly leafless**, rise well above tufted base of leaves.

LEAVES: Mostly at stem base, only one or two short leaves occur mid-stem; 2–9 mm wide; blades are either completely hairless or with a few fine, spreading hairs.

FLOWERS: In erect, flattened spike that is 5–10 cm long; each spikelet has many **bristle-like awns** that give it a feathery appearance early in spring and a stubbly appearance after fruits mature; upper spikelets soon fall off, **by mid-summer only bottom spikelets remain** (see photo).

HABITAT: Wet meadows, prairies, marshy areas and streambanks.

NATURAL HISTORY: Meadow barley is a widely adaptive species found in salt marshes, mountain meadows, open knolls, rocky ridges, and dry sage brush deserts, but generally where it is moist. It can form associations with one-sided sedge (*Carex unilateralis*) in the wetter sites of Willamette Valley wet prairies. The Oregon Natural Heritage Program is tracking this association due to its rapid decline in occurrence. The barbed awns of the spikelets hook onto the fur and hides of animals (and the clothing of people), which aids in seed dispersal.

SIMILAR SPECIES: The spikelets of blue-bunch wheatgrass (*Pseudoroegneria spicata*, also known as *Agropyron spicatum*, FACU-) are attached directly to the stem, alternating between sides, with irregular spaces between the spikelets. Other barley species (*Hordeum* spp.), sweet vernal grass (*Anthoxanthum odoratum*, p. 147) and rye grasses (*Elymus* spp.) can also be confused with meadow barley (consult technical manuals).

A: *Meadow barley* (Hordeum brachyantherum) *early in the growing season.*
B: *Meadow barley late in the season. The upper spikelets have fallen from the stems.*
C: *Meadow barley.*

SWEET VERNAL GRASS
Anthoxanthum odoratum

Poaceae, also called Gramineae (Grass Family)
INDICATOR STATUS: FACU

GROWTH HABIT: Perennial introduced grass; stems are sparsely leafy, with leaves at base and flower heads extending well above them.

LEAVES: Arise from base of stem, **flat, 30–60 cm long, 3–7 mm wide**, softly hairy on upper surface; **well-developed auricles, about 1 mm long**; occasional short leaf at mid-stem.

FLOWERS: In long (**2–9 cm**), **congested panicles** at stem tips; panicle is greenish to tawny in color, becomes straw colored by mid-summer; **spikelets of panicle are stacked in series of V-like clusters**.

HABITAT: Fields, pastures, lawns, wet meadows and wet prairies.

NATURAL HISTORY: Sweet vernal grass is native to Europe and is considered a weed in the Pacific Northwest. It can be an indicator of the transitions along wetland/upland boundaries or of marginal wetlands. Under fertile conditions sweet vernal grass grows larger and more robust than it does in poor soil or disturbed conditions.

SIMILAR SPECIES: Meadow foxtail (*Alopecurus pratensis*, p. 144) has similar leaves and habitats, but can be distinguished by its cylindrical panicles. Meadow barley (*Hordeum brachyantherum*, p. 146) is generally taller (40–80 cm), and its flowering spikes are 5–10 cm long, but they quickly fall apart after they mature. The leaves of meadow barley also lack auricles, while sweet vernal grass leaves have well-developed auricles.

NOTES: Sweet vernal grass gives off a sweet fragrance when it is crushed or burned.

A, B: *Sweet vernal grass* (Anthoxanthum odoratum).

CALIFORNIA OATGRASS
Danthonia californica

Poaceae, also called Gramineae (Grass Family)
INDICATOR STATUS: FACU

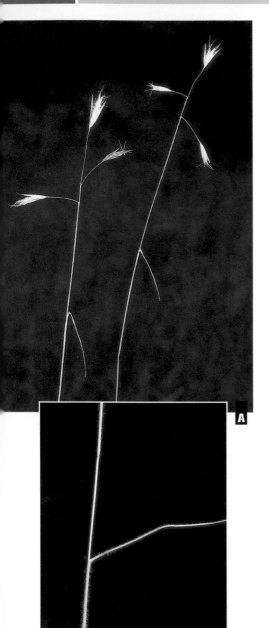

A

B

GROWTH HABIT: Perennial native grass; 30–80 cm tall; only 1–5 spikelets per plant; stems can be smooth and hairless, or have **1–2 mm long hairs**, especially **where leaf blades overlap stem at nodes** (see photo).

LEAVES: Sparse; found mostly on **lower stem; flat to somewhat rolled, 1.5–3 mm wide**.

FLOWERS: In loose, **open panicle of 1–5 spikelets**; panicle branches are thin and delicate; several flowers (florets) per spikelet.

HABITAT: Wet prairies and upland grasslands.

NATURAL HISTORY: California oatgrass forms seed in June and July. It is usually found in transitional areas along upland/wetland boundaries, and it is a component of wet prairie associations throughout the interior valleys of Oregon and Washington. This native grass is found on both sides of the Cascades, from the Pacific coast to the ponderosa pine forests of the eastern slopes.

SIMILAR SPECIES: California oatgrass is easy to recognize because it usually has only 1–5 spikelets per plant. California brome (*Bromus carinatus*, NOL) is similar, but it has 5–9 spikelets per plant (consult technical manuals).

A: *California oatgrass* (Danthonia californica).
B: *The stems and leaves of California oatgrass are usually hairy (although some plants have hairless stems and blades).*

TUFTED HAIRGRASS
Deschampsia cespitosa

Poaceae, also called Gramineae (Grass Family)
INDICATOR STATUS: FACW

GROWTH HABIT: Perennial native grass; 0.5–1.5 m tall; stems are erect, slender and densely tufted; **panicles are silky.**

LEAVES: Narrow, 8–20 cm long, 1.5–4 mm wide, often folded or rolled

FLOWERS: In open, **pyramidal panicles, 7–25 cm long,** that tend to droop but can be erect; **panicle branches are delicate, glistening,** often stiff and **grow upwards**; panicle is dark green and often purplish to brownish when in bloom, becomes silken straw color as it matures; **spikelets have soft hairs along main axis (rachilla) and at base of lemmas.**

HABITAT: Wet prairies, moist meadows and marshes.

NATURAL HISTORY: Tufted hairgrass is the dominant component in prairie bunchgrass wetlands and a dominant plant of the Willamette Valley wet prairie community. It is commonly associated with buttercups (*Ranunculus* spp.), Oregon saxifrage (*Saxifraga oregana*), northwest cinquefoil (*Potentilla gracilis*), asters (*Aster* spp.), camases (*Camassia* spp.), forget-me-nots (*Myosotis* spp.), rushes (*Juncus* spp.), and smaller sedges (*Carex* spp.). Tufted hairgrass is widely distributed in the Pacific Northwest. It is found from coastal salt marshes to subalpine wetlands. In floodplains and valleys tufted hairgrass plants tend to **form tussocks spaced 0.5–2 m apart**; at higher elevations they tend to form dense communities. The stems and inflorescences of tufted hairgrass remain upright through much of winter.

SIMILAR SPECIES: Annual hairgrass (*D. danthonioides*, FACW-) is a simple or somewhat tufted annual. It is smaller (5–50 cm tall) and has narrower leaves (1–1.5 mm wide). Slender hairgrass (*D. elongata*, FACW-) is a tufted perennial that is 25–80 cm tall and has shorter leaves (4–8 cm long and a mere 1 mm wide). Tufted hairgrass is much taller (to 150 cm) and larger than colonial bentgrass (*Agrostis capillaris*, p. 156) or creeping bentgrass (*A. stolonifera*, p. 156). Tufted hairgrass is also distinguished by the noticeable soft hairs in its spikelets.

Tufted hairgrass (Deschampsia cespitosa).

NOTES: Tufted hairgrass can be propagated from seed or by dividing the root masses, although older clumps can have dead centers.

TALL FESCUE · KENTUCKY FESCUE · ALTA FESCUE
Festuca arundinacea

Poaceae, also called Gramineae (Grass Family)
INDICATOR STATUS: FAC-

GROWTH HABIT: Perennial introduced grass; mostly **over 1 m tall**; stems (culms) are **erect, strongly tufted, hairless and smooth**, with **coarse, green or yellowish-green leaves**.

LEAVES: Stiff, **rough** on upper surface, **4–10 mm wide**, usually **flat but sometimes rolled inwards**; quickly become yellowish with age; **auricles at bases of blades have hairy edges** (ciliate).

FLOWERS: In **narrow panicles, 15–35 cm long**, at tops of stems; panicles are usually erect at first, but then tend to hang or droop to one side; seeds are small.

HABITAT: Wet fields, pastures and prairies and banks of ditches.

NATURAL HISTORY: Tall fescue forms seed from June to July. This introduced species is well established in wetlands and uplands of the Pacific Northwest from western Washington to northern California, and it can be an indicator of the transitional areas along wetland/upland boundaries or of marginal wetlands. The small seeds of tall fescue are eaten by songbirds, including golden-crowned and white-crowned sparrows.

A

B

C

D

SIMILAR SPECIES: The most diagnostic feature of tall fescue, even in winter, is the **fringe of hairs along the edges of the auricles** at base of blades. Red fescue (*F. rubra*, FAC+), a native species, sometimes occurs in wetlands from the coast to the mountains. It grows 20–80 cm tall (smaller than tall fescue) in loose clumps and lacks the hairy fringe on its auricles. Meadow fescue (*F. pratensis*, FACU+) has been widely introduced for forage. It is an upland species and its auricles are much smaller (less than 0.5 mm long) and not as well developed as those of tall fescue, and they lack the hairy fringe. Orchard grass (*Dactylis glomerata*, FACU), blue-joint (*Calamagrostis canadensis*, FACW), also called Canadian reed grass, and northwest mannagrass (*Glyceria occidentalis*, p. 80) can all be confused with tall fescue (and each other) when not in flower. However, the hairy auricles of tall fescue distinguish it from these other grasses (consult technical manuals).

A: *Tall fescue* (Festuca arundinacea). *The panicles are usually compressed along the stem.*

B: *A spikelet of tall fescue* (F. arundinacea).

C: *Red fescue* (F. rubra) *has full, spreading panicles.*

D: *Tall fescue* (F. arundinacea).

KENTUCKY BLUEGRASS
Poa pratensis

Poaceae, also called Gramineae (Grass Family)
INDICATOR STATUS: FAC

GROWTH HABIT: Perennial introduced grass; **30–150 cm tall**; grows from **strong underground stems (rhizomes)**; produces dense sod; **lower parts of stems (culms) are lax or curve downwards** at flattened, purplish, **rooting base**; upper stems are ascending and erect.

LEAVES: Flat or folded, 2–4 mm wide (appear narrower when folded), **up to 20 cm long, taper to keeled tip**; feel **rough** when pulled between fingers from tip to base.

FLOWERS: In slightly open, **pyramidal or oblong panicles**; panicles are erect, yellowish green or purplish, **up to 10–15 cm long**; **3–5 spikelets** clustered on each branch of panicle; **cobwebby hairs at base of each mature seed**.

HABITAT: Wet prairies, moist meadows and wooded wetlands.

NATURAL HISTORY: Kentucky bluegrass flowers from June to August, and it is very common in wetlands at moderate elevations. It was introduced as a forage plant and is widely used in seed mixes for lawns, where it escapes from cultivation and invades native plant communities, and it is now a serious pest. The seeds sometimes remain on the plants through winter and provide a food source for birds, such as wintering Canada geese and seed-eating songbirds.

SIMILAR SPECIES: The cobwebby hairs at the base of each mature seed distinguish Kentucky bluegrass from most other grasses. Annual bluegrass (*P. annua*, FAC) is widespread, especially in wetlands and lawns, has somewhat creeping stems and lacks the cobwebby hairs at the bases of the seeds. Fowl bluegrass (*P. palustris*, FAC), another European weed, is usually found in loose tufts in wetter habitats, and it has a larger panicle (10–30 cm long) with fine, spreading branches (usually in whorls of four or five). Rough bluegrass (*P. trivialis*, FACW) grows 40–100 cm tall from stolons (not rhizomes) in wetter habitats than Kentucky bluegrass.

A, B: *Kentucky bluegrass* (Poa pratensis).
C: *Annual bluegrass* (P. annua).

Poaceae, also called Gramineae (Grass Family)
INDICATOR STATUS: OBL

GROWTH HABIT: Annual native grass; to 1 m tall; light green, stout, erect; stems are robust, tall, cylindrical.

LEAVES: Flat, 5–10 mm wide.

FLOWERS: In 10–30 clustered, narrow spikes that are pressed close to top of stem (culm), each spike has two rows of 1-flowered spikelets on one side of its axis (rachis); spikelet rows are 10–30 cm long, with spaces between crowded groups of spikelets.

HABITAT: Wetter parts of bunchgrass prairies, moist meadows and vernal pools, marshy flats, ditches and mud of irrigated fields; also along edges of lakes, sloughs and ponds where it has been planted widely as wildlife food.

NATURAL HISTORY: American slough grass flowers mostly in June and July. It can be an indicator species for wetlands. The seeds provide food for waterfowl, seed-eating birds and small mammals.

SIMILAR SPECIES: American slough grass is distinguishable from all other grasses by the shape and neatly uniform 2-rowed arrangement of its flower spikelets.

A: *American slough grass* (Beckmannia syzigachne).

B: *American slough grass seed heads are compressed and jointed.*

C: *American slough grass.*

GROWTH HABIT: Annual introduced grass; to 50 cm tall (but usually much shorter in wetlands); stems are erect, **smooth and hairless**, and have inconspicuous nodes.

LEAVES: Sparse; found **mostly on lower stem**; 3–10 mm wide.

FLOWERS: In **open, pyramidal panicles** that are **5–15 cm long**; spikelets are broadly triangular, 2–5 mm long, jointed **(articulated) above glumes; glumes hang free at right angles to hair-like spikelet axis (rachilla)**, nod and tremble in wind, become purplish as spikelets mature and then become straw colored.

HABITAT: Wet prairies and disturbed areas.

NATURAL HISTORY: Little quaking grass forms seed from May through June. The seeds usually remain on plants in their dry form well into fall. Little quaking grass is a European grass thought to have been introduced for ornamental purposes. It has become well established in wet prairies and other upland areas in the Pacific Northwest. Where established, it is an early fire succession species and can dominate the ground cover in the first year after a fire in prairie wetlands.

SIMILAR SPECIES: Big quaking grass (*B. maxima*, NOL) is sometimes found east of the Cascades. Its spikelets are 10–19 mm wide. Rattlesnake grass (*Bromus briziformis*, NOL), another introduced species, resembles little quaking grass but it is much larger and it does not grow in wetlands or on the west side of the Cascades (consult technical manuals).

A, B, C: *Little quaking grass* (Briza minor).

Poaceae, also called Gramineae (Grass Family)

INDICATOR STATUS: FACU+

GROWTH HABIT: Annual native grass; 20–70 cm tall; hairy and wispy, hairs throughout are 2–4 mm long; stems (culms) are curved at base, branch freely.

LEAVES: 10–25 cm long, 5–15 mm wide, hairy on both sides; ligules are 1.5–2 mm long.

FLOWERS: In large, open, diffuse panicles, 10–30 cm long; spikelets are 1-flowered, 2–3.3 mm long, pointed at tips; panicle has purple hue in spring when anthers produce pollen; seeds are rounded, look like small, weightless beads.

HABITAT: Wet prairies, moist meadows and along banks of ditches, streams and ponds.

NATURAL HISTORY: Witchgrass is mostly considered to be a weed that grows in large irrigated tracts of disturbed land. In late summer the wispy, rounded panicles break from the stem and roll off with the wind like miniature tumbleweeds. This may be a seed dispersal mechanism. Witchgrass seeds provide food for waterfowl and songbirds.

SIMILAR SPECIES: Western witchgrass (*P. acuminatum*, also known as *P. occidentale*, FACW) is a perennial that grows in the wetter parts of similar habitats.

A: *Witchgrass* (Panicum capillare).
B: *Western witchgrass* (P. acuminatum).

GROWTH HABIT: Perennial introduced grass; sometimes 80 cm tall; **spreads via creeping, rooting stems** (stolons); erect **stems are in bunches of 3–5 and are jointed at nodes**; leaves and roots occasionally sprout from **noticeably bowed lower nodes.**

LEAVES: 2–5 mm wide, with **tiny hairs**; ligules at tops of leaf sheaths **are wider than long**, 0.5–2 mm long.

FLOWERS: In large, **open, pyramidal panicles, 7–15 cm long**, at stem tips; young panicles are erect and greenish with a purple cast; hardened base of lemma has **light beard of minute hairs**.

HABITAT: Varied, including wet meadows and prairies, fields, pastures, lawns and other grasslands and disturbed areas.

NATURAL HISTORY: Colonial bentgrass forms seed from June through August. The flower heads tend to remain long after the growing season is over. It was introduced from Europe and is now widespread in the West, though it is much more widespread in the eastern U.S. It is very common in a variety of wetland and upland habitats throughout the valleys and lowlands of the Pacific Northwest. It can be an indicator of the transitional areas along wetland/upland boundaries or of marginal wetlands. Colonial bentgrass spreads rapidly via its stolons and often dominates its habitats. Its seeds are a favorite food of voles, and are also eaten by nutria, deer and seed-eating birds.

SIMILAR SPECIES: Creeping bentgrass (*A. stolonifera*, formerly *A. alba* var. *stolonifera*, FAC) is a smaller wetland bentgrass that is usually no taller than 60 cm. It has a compressed panicle (up to 15 cm long), its branches are more ascending to erect and its 2–5-mm-long ligules are longer than wide. Leafy bentgrass (*A. diegoensis*, NOL) is usually 25–50 cm tall. It also has a more compressed panicle than colonial bentgrass and its sheath appendages (ligules) are 1–4 mm long. Redtop (*A. alba* var. *alba*, FAC) is a larger wetland bentgrass. It grows up to 120 cm tall and has narrower leaves, up to 4 mm wide. The bentgrasses differ from the bluegrasses (*Poa* spp., p. 152) in not

having keel-tipped leaves or cobwebby hairs at the base of each mature seed. Tufted hairgrass (*Deschampsia cespitosa*, p. 149) is much taller and larger than colonial bentgrass. It grows to 1.5 m tall, and has noticeable, soft hairs along the main spikelet axis (rachilla) and at the hardened bases of the lemmas. Consult technical manuals for more detailed descriptions.

NOTES: Colonial bentgrass is marketed in several horticultural forms as a common lawn grass.

A: *Colonial bentgrass* (Agrostis capillaris) *in a prairie habitat.*
B: *Colonial bentgrass* (A. capillaris).
C: *Creeping bentgrass* (A. stolonifera).
D: *Redtop* (A. alba *var.* alba).

DENSE SEDGE
Carex densa

Cyperaceae (Sedge Family)
INDICATOR STATUS: OBL

GROWTH HABIT: Perennial native grass-like; 30–100 cm tall; generally brown; **large, compact flower clusters** at stem tips; **stems are slightly triangular**, stout, **densely bunched**; lower stems surrounded by **translucent, papery sheaths** that are **puckered like miniature washboards**.

LEAVES: Alternate; scattered along lower stems; **flattened**; often longer than stems.

FLOWERS: In oblong to pyramidal inflorescence (2–10 cm long) of yellowish to brownish spikes ; 2–3 leaf-like bracts (each up to 5 cm long) at base of inflorescence.

HABITAT: Moist low ground or standing water of prairies, meadows, fields and pastures.

NATURAL HISTORY: Dense sedge forms seed from April through June. It is most common west of the Cascades in the valleys and on floodplains, but it is also found in higher-elevation wetlands and along the Columbia River. Dense sedge is an important food source for wildfowl, marsh birds, shorebirds, songbirds, beavers, muskrats, deer and small mammals.

SIMILAR SPECIES: Dense sedge is most easily confused with Cusick's sedge (*C. cusickii*, OBL), and it is very difficult to distinguish between the two without using a key. Generally, the leaves of Cusick's sedge are shorter than its flowering stems, while dense sedge's leaves tend to be longer than its flowering stems. Both sedges produce their flowers in spikes at the stem tips, but the flower spikes are more compressed in dense sedge, while the spikes of Cusick's sedge are less compact and tend to occur at irregular intervals. Dense sedge can also be distinguished by the puckered sheaths around its lower stems (see photo), while Cusick's sedge has reddish or coppery dots on its sheaths. Both hare's-foot sedge (*C. leporina*, also known as *C. ovalis*, FACW) and thick-headed sedge (*C. pachystachya*, FAC) are generally less than 80 cm tall, and their leaves are shorter than their stems. Also, their inflorescences are smaller than those of dense sedge. The inflorescence is 1.5–4 cm long in hare's-foot sedge and 1–2 cm long in thick-headed sedge. Awl-fruited sedge (*C. stipata*, p. 84) has more broadly triangular (and also narrowly winged) flower stalks (peduncles) and wider leaves (5–11 mm) than dense sedge.

A: *Dense sedge* (Carex densa).
B: *The characteristic puckering of the translucent sheaths that girdle the bases of dense sedge* (C. densa) *stems helps distinguish it from Cusick's sedge* (C. cusickii).
C: *Dense sedge* (C. densa).

Cyperaceae (Sedge Family)
INDICATOR STATUS: FACW

GROWTH HABIT: Perennial native grass-like; **30–110 cm tall**; **stems are usually coarse, stout,** fairly **leafy,** arising from dense clumps.
LEAVES: Flat, 2–5 mm wide; attach to stems at fairly wide intervals, not bunched at base; radiate outwards rather than rising from base of plant; sheaths at leaf bases are green all around without usual whitish translucence.

FLOWERS: In **compact** to somewhat **lengthened head,** 2–6 cm long, of **6–15 oblong spikes**; spikes are approximately **1 cm long, pale greenish to straw-colored,** attached directly to main stem (**sessile**), appear to be stacked one above another **in offset (90°) angle and crooked manner**.

HABITAT: Wet prairies and meadows, marshes and along banks of ditch.

NATURAL HISTORY: Green-sheathed sedge forms seed from May through July. It is common in the lowlands to moderate elevations of the mountains west of the Cascades and in the Columbia River Gorge.

SIMILAR SPECIES: Pointed-broom sedge (*C. scoparia*, FACW) has largely green sheaths (like green-sheathed sedge), but it has football-shaped, clustered flower spikes, while the spikes of green-sheathed sedge are clearly separated. Hare's-foot sedge (*C. leporina*, also known as *C. ovalis*, FACW) has brown flower spikes that have a 'shaggy' outline and are often tapered slightly at their bases. Hare's-foot sedge sometimes grows from stolons. Also, both pointed-broom sedge and hare's-foot sedge have curving or drooping flower stalks, while the flower stalks of green-sheathed sedge are nearly straight.

A

A, B: *Green-sheathed sedge* (Carex feta).

B

SLOUGH SEDGE
Carex obnupta

Cyperaceae (Sedge Family)
INDICATOR STATUS: OBL

A

GROWTH HABIT: Perennial native grass-like; **60–150 cm tall; densely clumped at base**, with long, thick, creeping underground stems (rhizomes); stems are coarse, stout and **conspicuously 3-sided; evergreen leaves** are wide and coarse with **sharp edges**.

LEAVES: Flat, dark green, 2–10 mm wide; surrounded at base by **reddish-brown, papery sheaths** that become **fibrous and shredded with age**.

FLOWERS: Typically in **4–8 long, spreading or drooping, brownish-purple spikes**, stalkless (sessile), or nearly so (**subsessile**) on very short stalks (peduncles); **upper 1–3 spikes are male; lower, female spikes are long, cylindrical**, somewhat narrow and pointed; bract attaches just below each spike, lowest bract is usually longest (much longer than spike), higher bracts are progressively smaller.

FRUITS: Round, glossy, very tough, brown perigynia; remain conspicuous all winter (distinctive feature of slough sedge).

HABITAT: Wet prairies and meadows, sloughs, shorelines and wooded wetlands.

NATURAL HISTORY: Slough sedge flowers from April through July. It is the most common of the few evergreen sedges in the valleys and floodplains of the Pacific Northwest, and also occurs in coastal wetlands and in forested wetlands at higher elevations. It is most often associated with Oregon ash (*Fraxinus latifolia*). Slough sedge often forms dense stands in standing water. New plants can grow 20–25 cm in their first year.

SIMILAR SPECIES: Lyngby's sedge (*C. lyngbyei*, OBL), which is more coastal than slough sedge, has its spikes (especially the upper ones) on longer and more erect flower stalks. The lower spikes tend to droop a bit. Columbia sedge (*C. aperta*, p. 82) is similar but has a copper-colored flowering cluster.

A: *Slough sedge* (Carex obnupta) *flower clusters.*
B: *An open slough sedge community at the edge of a shrub swamp wetland.*

B

Cyperaceae (Sedge Family)
INDICATOR STATUS: FACW

GROWTH HABIT: Perennial native grass-like; 30–100 cm tall; **stems are yellowish green,** densely **bunched; no underground stems (rhizomes)**; compact **flower head sits in axil of two or more unequal bracts.**

LEAVES: Alternate; arise from lower part of stem.

FLOWERS: In several **small spikes crowded together into compact head;** flower head sits in axis of two or more bracts; **lowest bract is 5–15 cm long (much longer than flower head),** and often **appears to be continuation of stem;** other bracts are shorter and grow more laterally from below flower head; flower head is greenish with light brown streaks throughout, becomes brown with age.

HABITAT: Common in wet fields and meadows that are semi-permanently saturated.

NATURAL HISTORY: One-sided sedge forms seed from May to July. It is common in mature Willamette Valley wet prairies and is also found along the Columbia River Gorge and over much of the West.

SIMILAR SPECIES: Slender-beaked sedge (*C. arthostachya*) looks very much the same as one-sided sedge, but the bracts under its flower heads are more bristle-like and are at a sharp angle to the stem, instead of seeming to be a continuation of it. Slender-beaked sedge is uncommon. Foothill sedge (*C. tumulicola,* NOL) is sometimes thought to look similar to one-sided sedge, but its flower head is not as compact and it is found in drier sites.

NOTES: The position of the flower head in the axil of unequal bracts gives rise to the 'one-sided' appearance of this sedge, which is the source of the common name and the species name *unilateralis*.

A, B: *One-sided sedge* (Carex unilateralis).

CREEPING SPIKE-RUSH
Eleocharis palustris (E. macrostachya)

Cyperaceae (Sedge Family)
INDICATOR STATUS: OBL

GROWTH HABIT: Perennial native emergent grass-like; **10–100 cm tall**; **stems are dark green**, erect, **round**, sometimes tufted, arise from small, branching underground stems (rhizomes).

LEAVES: Appears to lack leaves, but they are actually present, **modified as small, thin, dark reddish sheaths** that tightly **girdle bases of stems**.

FLOWERS: On lance-shaped, tapered spike, **5–25 mm long**, at tip of stem; **40–100 flowers per spike**, each enclosed by individual, **scaly bracts**.

HABITAT: Common in wet prairies, vernal pools, wet meadows, ditches, fields and pastures.

NATURAL HISTORY: Creeping spike-rush forms seed from May through August. It prefers semi-permanently saturated or flooded conditions and is most often found in standing water. The cluster of seeds in the spike provides food for various species of ducks and geese.

SIMILAR SPECIES: Needle spike-rush (*E. acicularis,* p. 85), the smallest of the three common spike-rushes in our area, is densely tufted. Ovate spike-rush (*E. ovata*, p. 86) is an annual of intermediate size.

NOTES: The name 'creeping spike-rush' refers to the plant's spreading rhizomes, not its stems. Creeping spike-rush is an erect plant and the tallest of the three most common spike-rushes of our region.

A, B, C: *Creeping spike-rush* (Eleocharis palustris).

TOAD RUSH
Juncus bufonius

Juncaceae (Rush Family)
INDICATOR STATUS: FACW+

GROWTH HABIT: Annual native grass-like; **5–30 cm tall**, height varies depending on site; bush-like, with one to many, thick, **short stems** and **dense branches**; **papery sheaths girdle stem bases**; fibrous roots; **stems become noticeably red** late in season or under dry, scorching conditions.

LEAVES: Linear, flat or often rolled along edges, usually **less than 1 mm wide**, 0.5–10 cm long.

FLOWERS: In panicles that can occur **along any part of stem**; **panicle branches are bow-like** (ascending but usually arched outwards); **flowers are tiny, sharp pointed, green with white edges**, attached directly to bowed branches.

HABITAT: Wet prairies and meadows, especially around springs and vernal pools, stream edges, roadsides, salt marshes and very disturbed sites, including wet pastures and wet edges of cultivated fields.

NATURAL HISTORY: Toad rush typically grows from June through August. It sometimes finds its way into gardens, where it is usually considered a weed.

SIMILAR SPECIES: Dwarf rush (*J. hemiendytus*, FACW+) is a tiny rush (just 1–2 cm tall) with a single flower at the tip of each stem.

A: *Toad rush* (Juncus bufonius). *Late in the season the stems become noticeably red.*
B: *A single toad rush flower.*
C: *A young toad rush plant.*

TAPER-TIPPED RUSH
Juncus acuminatus

Juncaceae (Rush Family)
INDICATOR STATUS: OBL

GROWTH HABIT: Perennial native grass-like; **20–80 cm tall**; **tufted**; **stems are round**, clustered on short underground stems (rhizomes), with flowers at branching tips.

LEAVES: Mostly at stem bases (**basal**), but there can be **1–3 stem leaves**, do not extend above flowers; **round, partitioned (septate)**; partitions in leaves can be felt by pinching leaf between thumb and forefinger, and running fingers up length of leaf.

FLOWERS: In **6–50 clusters at ends of loose, stiffly spreading branches** that are 3–10 cm high; **6–20 flowers per cluster**, tawny to greenish brown; flower clusters extend well above leaves and short (**2–15 cm long**), **leaf-like bract**.

HABITAT: Prairies, meadows, marshes and stream edges.

NATURAL HISTORY: Taper-tipped rush is especially evident from late June through August.

SIMILAR SPECIES: Taper-tipped rush, jointed rush (*J. articulatus*, p. 166) and Sierra rush (*J. nevadensis*, p. 167) can be easily confused with each other, and positive identification of these three species cannot be accomplished without technical keys. The following artificial differences help in making a preliminary, but not fully confident, determination of these species. The

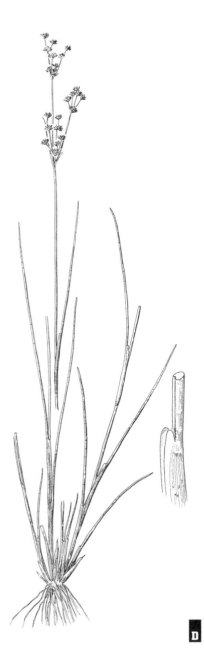

leaves of taper-tipped rush rise from the base of the plant and do not extend above the flower clusters, while those of jointed rush arise from different intervals along the stems and exceed the flower clusters. Jointed rush (15–50 cm tall) is usually shorter than taper-tipped rush. The flowers of jointed rush are typically deep brown. Sierra rush (10–70 cm tall) is about the same size as taper-tipped rush and its flowers also exceed the length of its leaves, but it generally has fewer flowers and they are purplish brown.

A, B, C: *Taper-tipped rush* (Juncus acuminatus).
D: *Taper-tipped rush, showing the sheath along the stem.*

JOINTED RUSH
Juncus articulatus

Juncaceae (Rush Family)
INDICATOR STATUS: OBL

GROWTH HABIT: Perennial native grass-like; **15–50 cm tall; tufted; stems are round**, clustered on short underground stems (rhizomes), with flowers at branching tips.

LEAVES: Round; do not extend above flowers; **1–3 leaves per stem**, with several shorter leaves at base of stem.

FLOWERS: In several **heads on loose, stiffly spreading branches** at stem tips; usually three flower heads on each 3–10 cm long branch; **leaf-like bract, 2–15 cm long,** positioned just below flower heads; flower heads extend well above leaves and bracts; **6–12 tawny to greenish-brown flowers per head**.

HABITAT: Prairies, meadows, marshes, and streambanks.

NATURAL HISTORY: Jointed rush is especially evident from late June through August.

SIMILAR SPECIES: See taper-tipped rush (*J. acuminatus*, p. 164) and Sierra rush (*J. nevadensis*, p. 167).

A, B: *Jointed rush* (Juncus articulatus).
C: *Jointed rush, showing the sheath along the stem.*

Juncaceae (Rush Family)
INDICATOR STATUS: FACW

GROWTH HABIT: Perennial native grass-like; **10–70 cm tall; strongly tufted,** strongly rhizomatous; **stems are round,** produce **leaves at various intervals; papery sheaths girdle stem bases and terminate in auricles (see illustration).**

LEAVES: 2–4 leaves per stem, round, less than 2 mm thick, conspicuously **cross-partitioned (septate);** do not extend past flower tops.

FLOWERS: In **1–4 clusters** (inflorescences) of flower heads **at stem tips, well above leaves;** less than 10 flowers per head; flowers are purplish brown, **turban-like, 3–3.5 mm wide;** short bracts attach below each cluster.

HABITAT: Wet prairies, moist meadows, wet streambanks and borders of lakes and ponds.

NATURAL HISTORY: Sierra rush occurs in four varieties. It is also found in coastal deflation plains and alkaline areas east of the Cascades. Sierra rush is reportedly good forage for domestic livestock.

SIMILAR SPECIES: See taper-tipped rush (*J. acuminatus,* p. 164), jointed rush (*J. articulatus,* p. 166), grass-leaf rush (*J. marginatus,* p. 168) and slender rush (*J. tenuis,* p. 169).

A: *Sierra rush* (Juncus nevadensis).
B: *Sierra rush rhizomes.*
C: *Sierra rush, showing the auricle at the top of the sheath along the stem.*

GRASS-LEAF RUSH
Juncus marginatus

Juncaceae (Rush Family)

INDICATOR STATUS: NOL (our region); FACW+ (nationally)

A

GROWTH HABIT: Perennial introduced grass-like; **15–70 cm tall; strongly tufted** from **knobby, nearly bulb-like underground stems (rhizomes); erect stems are round, slender, but thicker at base** (1–2.5 mm wide), with leaves at various intervals.

LEAVES: 3–4 alternate, flattened stem leaves; several soft, short leaves at stem bases; all leaves are shorter than the flower clusters.

FLOWERS: In usually open (but sometimes compact) **clusters (inflorescences), 1–10 cm long, at stem tips;** two to many small flower heads per cluster; **2–12 reddish-brown flowers per head; two narrow, leaf-like bracts** attach just below or with flower cluster, but usually do not surpass flowers.

HABITAT: Wet prairies, moist meadows and, especially, disturbed sites such as roadside ditches

NATURAL HISTORY: Grass-leaf rush was introduced to the Willamette Valley and has become a troublesome pest that invades habitats that are critical for native species. Its native range extends from Nova Scotia and Maine, southwards to Florida and westwards to Michigan, Missouri, Texas, Colorado, Arizona and California.

SIMILAR SPECIES: Grass-leaf rush looks very much like taper-tipped rush (*J. acuminatus*, p. 164), jointed rush (*J. articulatus*, p. 166) and Sierra rush (*J. nevadensis,* p. 167). However, these native species do not have the distinctive, knobby, bulb-like underground stems mentioned above, and their leaves are round rather than flattened.

NOTES: If you uproot a grass-leaf rush to check it for its distinctive, knobby, bulb-like underground stems, be sure to destroy the plant after verification to help slow its spread.

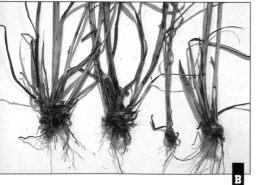

B

A: *Grass-leaf rush* (Juncus marginatus).
B: *Knobby, bulb-like, modified rhizomes arise from the roots of grass-leaf rush.*

GROWTH HABIT: Perennial native grass-like; **15–60 cm tall**; **stems are bright green, thin,** 0.5–2 mm wide, **bunched at base.**

LEAVES: Vestigial, inconspicuous, flat, only **1–3 mm wide**; sheath **at base of stem terminates in papery auricle.**

FLOWERS: In sparse to congested, **greenish to light brown panicle**, apparently from **side of stem**; two slender, **leaf-like bracts attach below panicle**; one bract is erect, looks like continuation of stem above panicle; other bract is shorter than panicle, extends from side of stem.

HABITAT: Wet meadows and prairies.

NATURAL HISTORY: Slender rush typically grows from June through September. It is usually associated with other grasses, sedges and rushes in the more exposed and wetter parts of meadows and prairies. It tolerates moderate disturbance.

SIMILAR SPECIES: Taper-tipped rush (*J. acuminatus*, p. 164) has noticeably cross-partitioned (septate) leaves, while slender rush does not. Soft rush (*J. effusus*, p. 170), which is dark green, and spreading rush (*J. patens*, p. 171), which is noticeably blue-green, both grow in tussocks. Slender rush, by contrast, is bright green and does not form tussocks. Slender rush and Sierra rush (*J. nevadensis*, p. 167) share the characteristic auricle at the top of the stem sheath, however the auricles of Sierra rush are more distinctly papery.

A: *Slender rush* (Juncus tenuis).
B: *A slender rush fruit and the characteristic auricle at the top of the sheath.*

SOFT RUSH • COMMON RUSH
Juncus effusus

Juncaceae (Rush Family)
INDICATOR STATUS: FACW

GROWTH HABIT: Perennial native emergent grass-like; 20–100 cm tall; **stems round, hollow, to 1 cm thick**, often brown or grayish late in season; strongly **tufted** in **crowded, thick tussocks**.

LEAVES: Round, **reduced to sheaths** that **surround stem bases**, though some may have vestigial, short, bristle-like blade.

FLOWERS: In **diffuse panicles up to 15 cm long**; appear to cascade from one side of stem because bract that attaches below panicle looks like continuation of stem.

HABITAT: Often in standing water of wet prairies, meadows, pastures and fields; also in shallow water at edges of pools, ponds and lakes.

NATURAL HISTORY: Soft rush typically grows from June to August. It is the most common rush in the coastal and interior wetlands and is a good indicator species for these habitats. It is not a major component of native wetland communities, however, and it often indicates disturbance. It usually forms dense stands in old wet pastures, nearly excluding all other plants. In drier habitats it grows in association with grasses and in wetter conditions it often occurs with buttercups.

SIMILAR SPECIES: Soft rush is an extremely variable species and has many forms and varieties. The flower head of *J. effusus* var. *pacifica* is more compact, while that of *J. effusus* var. *gracilis* is more open. Baltic rush (*J. balticus,* FACW+), which is also extremely variable, can be distinguished from soft rush only by technical details of the stamens (consult technical manuals). Spreading rush (*J. patens,* p. 171) also has very similar stems, leaves and flower clusters, but it is distinguished by its blue-green stems.

A: *Soft rush* (Juncus effusus) *flower clusters appear to cascade from the side of the stem.*
B: J. effusus *var.* gracilis.
C: J. effusus *var.* pacifica.

A

GROWTH HABIT: Perennial native grass-like; **15–100 cm tall**; grows in **tussocks; stems are blue-green, round**, slender, **often weak and bowed at base**, with **pointed tips**.

LEAVES: Vestigial, in form of loose, **brown sheaths** that surround stem bases, with **tiny, bristle-like blades**.

FLOWERS: In small, short, **congested, branched clusters**, apparently from **side of stem** about two-thirds of way up; what appears as **continuation of stem above flowering cluster is actually an erect, sharp-pointed bract**, 10–35 cm long.

HABITAT: Wet prairies and meadows, stream edges and moist to wet woodlands.

NATURAL HISTORY: Spreading rush typically grows from July through August.

SIMILAR SPECIES: Spreading rush is generally similar to soft rush (*J. effusus*, p. 170), but soft rush is dark green and is a much more robust plant.

B

A: *Spreading rush* (Juncus patens) *in a typical habitat.*

B: *Spreading rush flower clusters seem to arise from the sides of the stems.*

Shrub swamps and brush prairies are wetlands dominated by shrubs. They occur on floodplains and gravel bars and in stream channels, and most tolerate variable water flow. Community structure ranges from scattered shrubs with intervening herbaceous plants to dense and impenetrable stands of shrubs and trees.

Brush prairies occupy former sites of tufted hairgrass prairie and develop successionally in the absence of fire. Nootka rose (*Rosa nutkana*) and tufted hairgrass (*Deschampsia cespitosa*, chapter 3) form a primary community, often with small openings of herbaceous prairie plants.

Rose/Douglas' spiraea (Rosa sp./Spiraea douglasii) communities may contain one or a combination of Nootka rose (R. nutkana), clustered rose (R. pisocarpa) or sweetbrier (R. eglanteria).

Associated species include serviceberry (*Amelanchier alnifolia*), Douglas' hawthorn (*Crataegus douglasii*) and Douglas' spiraea (*Spiraea douglasii*). In many areas weedy shrubs, such as English hawthorn (*Crataegus monogyna*), naturalized apples and crabapples (*Pyrus* spp.) and sweetbrier (*Rosa eglanteria*), form dense thickets. Dwarf huckleberry (*Vaccinium caespitosum*), known from only two prairie sites, is probably a rare relict of the pre-settlement vegetation.

Shrub swamp communities along watercourses are highly variable and difficult to classify. Many contain various mixtures of the same species, with or without a partial tree canopy. Red-osier dogwood (*Cornus sericea*, chapter 5) and Douglas' spiraea form two common watercourse communities that are variations of woodland communities with similar components. Douglas' spiraea is one of the few native plants that benefits from disturbances. It often invades flooded pastures and areas with disturbed drainage.

Yellow willow (*Salix rigida*), northwest willow (*S. sessilifolia*) and Columbia River willow (*S. fluviatilis*) form often impenetrable stands on gravel and sand bars along major rivers. Some of these stands have been aided by flood control—they have developed on alluvial flats

Brush prairie communities are mixtures of woody, herbaceous and grass species. Willows (Salix spp.), roses (Rosa spp.), serviceberry (Amelanchier alnifolia), hawthorns (Crataegus spp.) and Pacific crabapple (Pyrus fusca) are among the most common woody species.

Riparian willow communities form on gravel bars associated with large streams, rivers and floodplains.

that were formerly stripped clean by annual flooding.

Along the lower Willamette and Columbia rivers, Pacific willow (*Salix lucida* ssp. *lasiandra*) and reed canary-grass (*Phalaris arundinacea*, chapter 3) form a common community type. Much of the native understory was destroyed by livestock grazing, drainage, and mosquito and flood control measures, but pre-settlement components probably included Columbia sedge (*Carex aperta*, chapter 2), green-sheathed sedge (*Carex feta*, chapter 3), woolly sedge (*Carex pellita*, chapter 5), retrorse sedge (*Carex retrorsa*) and stinging nettle (*Urtica dioica*, chapter 5).

In pre-settlement times, willow swamps were second in our area only to wet prairies in a landscape that was dramatically different from what we see today. Most of the early explorers wrote about how much water and swamp there was, and how many beavers there were. Willows were the beavers' principal food. Today's landscape supports only one tenth the number of beavers that were here in 1800. Hundreds of thousands of hectares of bottomland soils, now fertile agricultural fields, are in part the legacy of the beavers.

Floodplain willow swamps typically contain stands of reed canary-grass (Phalaris arundinacea).

Geyer's willow (*Salix geyeriana*) and Hooker's willow (*S. hookeriana*) form a rare community on cold, organic soils in the lower Willamette Valley. When the first settlers arrived, this community occupied some 4,000 hectares of bottomland. All but one percent of it was cleared for onion fields and pasture. Associated species included bog birch (*Betula glandulosa*), Labrador tea (*Ledum glandulosum*), yellow marsh marigold (*Caltha asarifolia*), water sedge (*Carex aquatilis* var. *dives*), bogbean (*Menyanthes trifoliata*) and marsh cinquefoil (*Potentilla palustris*, chapter 3), which are all now restricted to montane or coastal peatlands.

Willow swamps include associations of nodding beggarticks (Bidens cernua) *and waterpepper* (Polygonum hydropiperoides).

Onagraceae (Evening-primrose Family)
INDICATOR STATUS: FACW-

GROWTH HABIT: Perennial native forb; typically **30–100 cm tall**; spreads by short underground stems (rhizomes); main stems branch in upper half, especially in flower clusters.

LEAVES: Opposite, densely arranged on stem, strongly veined, **glossy green, lance- to egg-lance-shaped, 1–15 cm long**, stalkless (**sessile**) or nearly so (**subsessile**), appear to attach directly to stems; **edges minutely toothed**.

FLOWERS: In loose, **compound clusters (inflorescences)**; **pink to rose-purple; four petals, 5–14 mm long**, light pink, **notched at edges**; fruit forms below flower (inferior ovary).

HABITAT: Shallow, fresh water, including marshes, shrub swamps and wet meadows, especially in disturbed habitats.

NATURAL HISTORY: Watson's willow-herb is often found in the understory of cattail (*Typha latifolia*), sedges (*Carex* spp.) and rushes (*Juncus* spp.). It is native to western North America, but it has spread over much of the United States and has invaded Europe.

SIMILAR SPECIES: Smooth willow-herb (*E. glaberrimum* ssp. *glaberrimum*, FACW) has small (1–7 cm long), narrow, opposite leaves that clasp the stem and appear to attach directly to it. Common willow-herb (*E. ciliatum* ssp. *glandulosum*, also known as *E. glandulosum*, FACW-) has fleshy, rosebud-like offshoots (turions) on its roots or rhizomes at maturity. You have to uproot a plant to see this feature. Dense spike-primrose (*E. densiflorum*, p. 118) has opposite, hairy leaves (except those near the stem base), and its stems peel towards the base. The leaves of purple loosestrife (*Lythrum salicaria*, p. 75) attach directly to the stem and have grayish hairs on their upper surfaces. They can appear to be in whorls around the stem, but actually they are opposite. The flowers of purple loosestrife are densely crowded in the upper leaf axils, in an interrupted pattern. Hyssop loosestrife (*Lythrum hyssopifolium*, p. 75) is a pale waxy-bluish green annual with both opposite and alternate leaves (in threes). Also, both species of loosestrife are much larger than Watson's willow-herb.

NOTES: The willow-herbs belong to the same genus as fireweed (*E. angustifolium*).

A, B: *Watson's willow-herb* (Epilobium ciliatum *ssp.* watsonii).

COW PARSNIP
Heracleum lanatum

Apiaceae, also called Umbelliferae (Carrot Family)
INDICATOR STATUS: FAC+

Cow parsnip (Heracleum lanatum). *The leaves and flowers emerge from sheaths along the stem.*

GROWTH HABIT: Perennial/biennial native forb; **1–3 m tall**; each plant has one **large hollow stem** with **woolly hairs** that are thicker near the leaves and flowers.

LEAVES: Large, **divided into three deeply lobed leaflets**; each leaflet **twice more divided**, with shallow indentations and coarse serrations along its edges; **leaflets are 10–40 cm long and about as wide; leaflet undersides are covered with matted or shaggy hairs**; main leaf stalks (petioles) are 10–40 cm long; **upper leaf stalks are skirted by inflated, husk-like sheaths** that enclose leaves before they open and remain at leaf bases once leaves have expanded.

FLOWERS: In **compound umbel, 10–20 cm across**, on stout, woolly or long-hairy stalk (peduncle), 5–20 cm long, at stem tip; 15–30 secondary umbels (umbellets), each on stalk (ray) up to 10 cm long; individual **flowers are small, white**; flower stalks (pedicels) are 8–20 mm long; second, smaller umbel will sometimes occur lower on stem.

HABITAT: Streambanks, wet meadows and other moist low ground.

NATURAL HISTORY: Cow parsnip blooms from early June through August. It is the only species of its genus that is native to North America. It is not known whether cow parsnip seeds are eaten by wildlife, but its plentiful nectar attracts insects. The sheaths on the leaf stalks provide a refuge for slugs.

SIMILAR SPECIES: Giant hogweed (*H. mantegazzianum*, NOL) is a huge (1.5–4.5 m tall), related species that is grown in gardens and has become naturalized in some areas. Coltsfoot (*Petasites frigidus* var. *palmatus*, FACW-) can grow to 60 cm tall and has white flowers that bloom in late winter or early spring. It is a member of the sunflower family, with a very different structure to its flowers, which are borne in a feathery head rather than an umbel. Also, coltsfoot flowers are often pink-tinged while cow parsnip flowers are white only.

NOTES: Cow parsnip is one of the largest herbaceous wetland plants. It is typically rank smelling, and some people can get contact dermatitis from it.

SERVICEBERRY
Amelanchier alnifolia

Rosaceae (Rose Family)
INDICATOR STATUS: FACU

GROWTH HABIT: Native shrub; **1–5 m tall**; large **shrub or small tree** (sometimes); new stems are reddish brown and smooth; new bark can have thin layer of woolly hairs; mature stems, twigs and branches are hairless and eventually turn gray.

LEAVES: Deep green; oval to oblong, **2–4 cm long**; leaf edges are **toothed towards leaf tip**; leaf stalks (petioles) are slender, 10–25 mm long.

FLOWERS: Clustered at ends of branches; **pure white; petals are slender, long and narrow, appear to spread and twist**.

FRUITS: Fleshy, blue pomes; eventually turn black; several seeds in central core (like an apple).

HABITAT: Shrub swamps, streambanks and wooded wetlands; also invades wet prairies and many upland sites.

NATURAL HISTORY: Serviceberry blooms during May and June. Its fruits provide food for birds and, at higher elevations, bears.

SIMILAR SPECIES: Indian plum (*Oemleria cerasiformis*, FACU) leaves are long and have rather pointed tips, while serviceberry leaves are short, broad and blunt. Serviceberry leaves look a little like the leaves of wild roses (*Rosa* spp., p. 182), but wild rose leaves are pinnately compound and the leaflets have saw-toothed (serrate) edges. Serviceberry leaves are simple and toothed only towards the tip. Snowberry (*Symphoricarpos albus*, p. 210) leaves are usually smaller than serviceberry leaves, and they are opposite, lighter green and slightly toothed along their edges.

NOTES: Serviceberries were a staple food for Native Americans, and they were apparently eaten by Lewis and Clark. They are good in jams and pies.

A, B: *Serviceberry* (Amelanchier alnifolia).

DOUGLAS' HAWTHORN • BLACK HAWTHORN
Crataegus douglasii

Rosaceae (Rose Family)
INDICATOR STATUS: FAC

GROWTH HABIT: Native tree; **1–4 m tall**; **large shrub or** (usually) **small tree**; **stems have spines** that can be slightly **curved or straight**; older spines are 1–2 cm long.

LEAVES: **3–6 cm long, half as wide**; leaf stalks (petioles) are short; **leaf edges deeply toothed from tips to middle of blade** (where dentation stops abruptly), then inconspicuously saw-toothed (serrate) from middle of blade to leaf stalk.

FLOWERS: In clusters above leaves at ends of branches; **flowers are white and have hypanthiums** (floral cups); petals are nearly **round, 5–7 mm long**; generally **five styles** and **10 stamens**.

FRUITS: **Purplish-black drupes**; shaped like small cherries and have a central pit; hang in clusters; persist through winter.

HABITAT: Shrub swamps, watercourses, and wooded areas bordering and invading wet prairies.

NATURAL HISTORY: Douglas' hawthorn blooms during May and June. Its fruits are an important source of food through winter for fox sparrows and cedar waxwings. The thorns, dense branches and thick foliage make Douglas' hawthorn a favorite, safe nesting site for many birds. Douglas' hawthorn is known to host commercial fruit tree pests.

SIMILAR SPECIES: A variety of Douglas' hawthorn called Suksdorf's hawthorn (*C. douglasii* var. *suksdorfii*, NOL) has more oblong, elliptical, unlobed leaves with slight serrations around their edges. Its leaves are lighter green, smooth and shiny, its spines are shorter (8–12 mm long), and its flowers have five styles and usually have 20 stamens. Two widespread, introduced species, *C. monogyna* (FACU+) and *C. oxyacantha* (NOL), which are both called English hawthorn, often escape from cultivation and can both be mistaken for Douglas' hawthorn. They can be distinguished from Douglas' hawthorn by their leaves, which are lobed like oak leaves (see photos) and by their flowers— *C. monogyna* flowers have one style and *C. oxyacantha* flowers have two styles. Also see Pacific crabapple (*Pyrus fusca*, p. 180).

C

D

E

A: *Douglas' hawthorn* (Crataegus douglasii). *Note the large thorn on the stem.*

B: *Suksdorf's hawthorn* (C. douglasii *var.* suksdorfii).

C: *English hawthorn* (C. monogyna) *in full bloom. Note the deeply divided lobes around the leaf edges.*

D: *Douglas' hawthorn* (C. douglasii).

E: *English hawthorn* (C. monogyna).

PACIFIC CRABAPPLE • WESTERN CRABAPPLE
Pyrus fusca (Malus fusca)

Rosaceae (Rose Family)
INDICATOR STATUS: FAC+

A

GROWTH HABIT: Native tree; **5–12 m tall**; **shrub to small tree**; young twigs are covered with tiny white or gray hairs; twigs and branches **appear to have thorns**, but these are **actually spurs**, on which flowers and fruits are produced.

LEAVES: Similar to domestic apple leaves; **light green, lance-shaped to oval-oblong**, 4–40 cm long; **edges saw-toothed (serrate) and slightly curled**; unlike domestic apple, edges are often **lobed near tip**.

FLOWERS: In clusters; petals are oblong-oval to obovate, 9–14 mm long, **white to pink**.

FRUITS: In clusters; **polished yellow to purplish-red pomes, 10–16 mm long**.

HABITAT: Shrubby and wooded swamps and low, wet woods.

NATURAL HISTORY: Pacific crabapple blooms from late April to June. It is often found growing in association with Oregon ash (*Fraxinus latifolia*) in Willamette Valley wetlands. Both tree species often have a frosty appearance due to a covering of lichens. Pacific crabapples are a favorite food of deer, elk and bears.

SIMILAR SPECIES: Several domesticated fruit trees commonly escape cultivation and are well established in valleys and floodplains. Wild pear (*Pyrus communis*, NOL), wild apple (*Pyrus malus*, FACW) and wild plums and cherries (*Prunus* spp.) can be confused with Pacific crabapple. The hawthorns, especially Douglas' hawthorn (*Crataegus douglasii*, p. 178), can also be confused with Pacific crabapple.

B

C

A: *Pacific crabapple* (Pyrus fusca).
B: *Domestic apples* (P. malus) *often escape cultivation and become established in shrub swamps.*
C: *Pacific crabapple* (P. fusca).

DOUGLAS' SPIRAEA • HARDHACK
Spiraea douglasii

Rosaceae (Rose Family)
INDICATOR STATUS: FACW

A

GROWTH HABIT: Native **shrub**; **1–2 m tall**; stems are woody and slender, branch freely, can be so numerous that plant is almost impenetrable, especially in marshy areas or around lakes.

LEAVES: Oval, 3–10 cm long; leaf stalks (petioles) are short; **upper leaf surface is dark green**; underside is covered with fine hairs, or is hairless; **leaf edges are notched or saw-toothed (serrate) along upper half**.

FLOWERS: Clustered in showy, plume-like panicles at tops of stems; each panicle is oblong to conical, 6–20 cm long; flowers have five tiny, pale to deep rose-pink petals, and appear hairy due to many red stamens that extend beyond petals, dry brown after only one day or so.

HABITAT: Streambanks, lakeshores, roadside ditches, swamps and bogs at most elevations west of Cascades.

NATURAL HISTORY: Douglas' spiraea blooms from July to August. It roots easily from cuttings, can tolerate flooding for long periods of time, and is well adapted to grow in disturbed wetlands.

SIMILAR SPECIES: From a distance Douglas' spiraea can be confused with fireweed (*Epilobium angustifolium*, p. 75) or one of the

B

loosestrifes (*Lythrum* spp., p. 75), but the woody stems and the flowers and leaves of Douglas' spiraea are distinctive when seen up close.

A: *Douglas' spiraea* (Spiraea douglasii) *with Nootka rose* (Rosa nutkana). *Douglas' spiraea flowers soon turn brown.*
B: *Douglas' spiraea* (S. douglasii).

GROWTH HABIT: Native **shrub**; **1–2 m tall**; stems are wiry twigs, usually armed with many, **large, paired thorns**, or **can be nearly thornless**.

LEAVES: Alternate; **compound, with 5–7 leaflets**; leaflets are **elliptical to oval**, 1–7 cm long, 0.7–4.5 cm wide, with **shallow teeth along edges**.

FLOWERS: Nearly always occur **singly** at ends of new side branches; **large (2.5–4 cm across)**, attractive, **light pink to deep rose**.

FRUITS: Fleshy, round, orange-red 'hip,' 1–1.5 cm long.

HABITAT: Shrub swamps, thickets at edges of forested wetlands, lakeshores and streambanks.

NATURAL HISTORY: Nootka rose begins blooming in May and the fruits mature by July. It is common in both the interior lowland valleys and at higher elevations, and it often invades and becomes thickly established in prairie wetlands that have not been routinely

burned. The hips, which remain on the plants through winter, provide food for wildlife when other sources are limited. Wild rose thickets are good cover and nesting areas for game-birds, songbirds and small mammals.

SIMILAR SPECIES: Clustered rose (*R. pisocarpa*, FAC) is found in the same habitats as Nootka rose and often occurs with it, but clustered rose is confined to lower elevations. Its stems can often be almost thornless (like Nootka rose), but its leaves, flowers, thorns and fruits are all smaller than those of Nootka rose. The flowers and fruits of clustered rose are clustered in spreading groups of three to more than 20 at the ends of the youngest branches. Clustered rose hips are small (6–12 mm long), purplish and pear-shaped. The leaves of clustered rose are hairless and glandless, but there are stalked glands at the base of the sepals at the ends of the hips. Sweetbrier (*R. eglanteria*, FACW), which can often be found growing with Nootka rose and clustered rose, gives off a sweet fragrance when its leaves are crushed. It is non-native, has few olive-green prickles and nearly always has galls (see photo). In our region, sweetbrier is the only rose to get galls.

The following is a good guide for distinguishing between the three wetland roses:

Nootka rose: Scattered, stalked glands along the underside of the main leaf axis (rachis); gland-tipped teeth along the leaf edges.

Clustered rose: No glands on the leaf undersides or along the leaf edges; stalked glands at the base of the sepals on the fruit.

Sweetbrier: The leaves give off a sweet fragrance when they are crushed; this aroma is sometimes very noticeable in the air on warm, summer days.

NOTES: Rose hips can be made into a preserve or jam, and are a rich source of vitamin C. Wild roses root easily from cuttings.

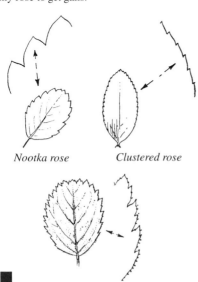

Nootka rose   *Clustered rose*

Sweetbrier

A: *Nootka rose* (Rosa nutkana).

B: *Clustered rose* (R. pisocarpa). *The gland-tipped hairs aid in identification.*

C: *Sweetbrier* (R. eglanteria) *often has galls (fringed, tangled growths that form in response to the larvae of cynipid gall wasps).*

D: *A 'hip' of Nootka rose* (R. nutkana).

E: *Leaflets of Nootka rose, clustered rose and sweetbrier, showing the leaf edges.*

SCOULER'S WILLOW
Salix scouleriana

Salicaceae (Willow Family)
INDICATOR STATUS: FAC

A

GROWTH HABIT: Native tree; **2–12 m tall**; **shrub** or more often **small tree**; trunk often 10 cm thick, can grow to 40 cm; new stems and twigs have short, straight hairs that are closely pressed along stem; **bark often striped in 2-tone pattern of greenish and reddish brown**, and can have **skunk-like odor**.

LEAVES: Oval to lance-shaped, broadest near tip; **upper leaf surfaces are green, glossy**, become hairless with age (except for **sparse hairs along midvein**); **leaf undersides are waxy with dense, white**, often **rusty hairs**; leaf edges occasionally have fine teeth; leaf stalks (petioles) are 5–10 mm long.

FLOWERS: In catkins on short (seldom over 1.5 cm) stalks (peduncles); **male (staminate) catkins are 2–4 cm long**, soon fall off; **female (pistillate) catkins are 2.5–6 cm long**; tiny, **dark brown to black bracts** attach just below cottony seeds of female catkins.

HABITAT: Shrub swamps, moist woods and clearings, streambanks and lakeshores.

NATURAL HISTORY: Scouler's willow develops catkins in early spring, usually before the new leaves appear. Of all the willows in our area, Scouler's willow tolerates the driest conditions; it can grow far from water, even in upland forests. Willows are browsed by deer, rabbits and other animals, especially in spring when the tender shoots are the first foliage to appear.

SIMILAR SPECIES: See the other willows, especially Hooker's willow (*S hookeriana*, p. 185) and Sitka willow (*S. sitchensis*, p. 186).

NOTES: Native Americans in the Pacific Northwest used willow bark for many things, including fish nets, baskets and twine. Willow bark contains substances that, like aspirin, are anti-inflammatory.

B

A: *Scouler's willow* (Salix scouleriana). *The new twigs often have a 2-toned, greenish- or reddish-brown mottling.*

B: *Scouler's willow leaves have shiny upper surfaces and gray, velvety undersides.*

Salicaceae (Willow Family)
INDICATOR STATUS: FACW-

GROWTH HABIT: Native shrub; **3–6 m tall**; **shrub** or more often **small tree**; twigs are brownish, slightly soft-hairy as they emerge, soon become hairless, smooth and shiny.

LEAVES: **Broadly elliptical, up to 4 times as long as wide**, with either rounded or slightly pointed tips; **leaf undersides** have **waxy, grayish blue coating** (glaucous) and **shaggy hairs**; **hairs are white to white-rusty when leaves are new** and not fully expanded; **upper leaf surfaces are highly glossy**, with densely **matted hairs**; leaf edges are loosely toothed; well-developed, **paired, leaf- or collar-like stipules** at bases of leaf stalks (petioles).

FLOWERS: In **catkins** on short (up to 1 cm long) stalks (peduncles); **male (staminate) catkins are stout**, mostly 3–4 cm long and 2–2.5 cm wide; **female (pistillate) catkins are 4–12 cm long**, have dense, **cottony seeds**; **brown to black, persistent scales** attach below each flower; **leaf-like bracts**, small and inconspicuous or up to 3 cm long, in pairs at base of each catkin stalk.

HABITAT: Especially common in coastal habitats, but also occurs inland in shrub swamps, on streambanks and along moist roadsides.

NATURAL HISTORY: Hooker's willow produces catkins in March and April. It is widespread in the western interior valleys of the Pacific Northwest and extends as far north as Mt. Rainier, where it occurs at moderate elevations.

SIMILAR SPECIES: Leaf shape and other characteristics vary considerably among the individuals of Hooker's willow, making it difficult to identify. See other willows, especially Scouler's willow (*S. scouleriana*, p. 184), and Sitka willow (*S. sitchensis*, p. 186).

NOTES: Piper's willow (*S. piperi*) was formerly considered a separate species, but it is now officially submerged into *Salix hookeriana*.

A: *The female catkins of Hooker's willow* (Salix hookeriana) *produce cottony seeds.*
B: *The male catkins of Hooker's willow appear in early spring.*
C: *Hooker's willow leaves have very glossy upper surfaces.*

SITKA WILLOW
Salix sitchensis

Salicaceae (Willow Family)
INDICATOR STATUS: FACW

GROWTH HABIT: Native **shrub**; **2–6 m tall**; many-stemmed, with dark, **densely velvety twigs**; bark of older stems is **smooth and gray**.
LEAVES: Narrowly oval to lance-shaped, **widest near tips**, 4–9 cm long, 1.5–3.5 cm wide; **dark green and shiny on top; dense, silvery-white hairs below** that resemble **crushed velvet** and have characteristic sheen when blown by wind; leaf stalks (petioles) are 5–15 mm long; **pair of leaf-like stipules** can be found at bases of some leaves, and are especially well developed on vigorous, new shoots.
FLOWERS: In catkins on 1-cm-long stalks (peduncles); **male (staminate) catkins are 2.5–5 cm long**, about 1–1.5 cm wide; **female (pistillate) catkins are 3–8 cm long**, with **brown to black bracts** below each flower; a **pair of leafy bracts** (approximately 2 cm long) is **attached at base of catkin stalk**.
HABITAT: Shrub swamps, marshes, boggy places and streambanks.
NATURAL HISTORY: Sitka willow produces catkins before its leaves bud. It is common throughout most Pacific Northwest wetlands, especially in the valleys and floodplains west of the Cascades. It also occurs at moderate elevations in the Cascades, the Wallowa Mountains and eastern Washington, and it ranges north into British Columbia and south to San Luis Obispo County in California.
SIMILAR SPECIES: See other willows, especially Hooker's willow (*S. hookeriana*, p. 185) and Scouler's willow (*S. scouleriana*, p. 184). Some forms of Sitka willow and Scouler's willow are impossible to distinguish without closely examining the flowers. In general, however, the **'crushed velvet' hairiness on the undersides of the leaves** is the best diagnostic feature for Sitka willow.

A: *Sitka willow* (Salix sitchensis). *The female catkins are 3–8 cm long.*
B: *Sitka willow leaves have rust-colored veins on their undersides.*
C: *Sitka willow.*

Salicaceae (Willow Family)
INDICATOR STATUS: FACW+

GROWTH HABIT: Native shrub; **large shrub, 3–6 m tall**, or **small tree, to 10 m tall**; shrubs produce many coarse stems, new twigs are minutely hairy, glossy and red to olive; as tree, trunk will grow to 30 cm thick, **bark is smooth and gray when young, becoming dark gray and fissured when older**.

LEAVES: Most are **long and narrow**, 5–15 cm long, 1–3 cm wide, with **tapered tip (acuminate)**, similar in shape to leaves of peach tree; leaves on vigorous, new shoots are particularly large, up to 25 cm long and 5 cm wide; **leaves on flowering branches are smaller**, relatively **broader and blunter**; all have few if any hairs when young, **hairless when mature; leaf edges have fine, close serrations**; two to several **long, wart-like glands** on leaf stalk (petiole) at base of blade; pair of well-developed **stipules at base of leaf stalk**, look like neat little leaf collars and eventually fall off.

A

FLOWERS: In **long, thick catkins; male (staminate) catkins are 2–7 cm long**, 1–1.5 cm wide; **female (pistillate) catkins are 3–12 cm long; yellow, minutely hairy, deciduous bracts** attach below each flower.

HABITAT: Streambanks, swamps and marshy thickets.

NATURAL HISTORY: Pacific willow catkins develop after the leaves have expanded in spring. Pacific willow occurs in most of western North America from California to Alaska, including both sides of the Cascades, the Rocky Mountain states, Alberta and British Columbia. Willows are pioneer species that become established on sand and gravel bars as well as other places where soil development is poor. Their wood and branch attachments are often brittle, and stems break easily. This brittle nature may be a useful dispersal mechanism used for invading new habitat. Pacific willow is a major food plant for beavers.

SIMILAR SPECIES: The **glands at the leaf bases** are a distinctive feature of Pacific willow. Yellow willow (*S. rigida*, OBL) usually grows as a many-stemmed, coarse shrub, 2–4 m tall, but it is sometimes up to 9 m tall and tree-like, with a trunk to 10 cm thick. Yellow willow has slightly broader leaves than Pacific willow, especially at

B

the base of the blade, and its catkins are slightly smaller, but the most distinctive difference is that yellow willow does not have glands on its leaf stalks. Yellow willow is generally uncommon, except on gravel and sandbars along major rivers, where it often forms impenetrable stands with Columbia River willow (*S. fluviatilis*, p. 190) and northwest willow (*S. sessilifolia*, p. 189).

NOTES: 'Nehalem' Pacific willow is a cultivar of Pacific willow that is used to stabilize streambanks and is planted along the shores of lakes and man-made reservoirs.

A: *Pacific willow* (Salix lucida *ssp.* lasiandra). *Female catkins.*
B: *Male Pacific willow catkins.*

GEYER'S WILLOW
Salix geyeriana

Salicaceae (Willow Family)
INDICATOR STATUS: FACW+

A: *Geyer's willow* (Salix geyeriana). *The female catkins are short and stubby.*

B: *Geyer's willow at the edge of a prairie wetland.*

GROWTH HABIT: Native **shrub**; **4–6 m tall**; **short, narrow leaves, stubby catkins**, and slender shoots and twigs give plants dainty character even when they grow quite tall; twigs are yellowish to brownish; young stems and twigs are densely covered with cottony hairs; older twigs become smooth, hairless and glossy.

LEAVES: Narrowly **lance-shaped**, 3.2–7.4 cm long, 6–15 mm wide; **tip tapered**; **base wedge-shaped**; new, unfolding leaves are silky, clothed with dense hairs that are pressed flat on both sides; only upper surfaces remain sparsely hairy on mature leaves; **hairs are white to slightly rust colored**; **leaf edges are shallowly toothed**; leaf stalks (petioles) are 3–10 mm long.

FLOWERS: In catkins; **male (staminate) catkins are small**, 7–15 mm long, on 1–2.8 cm long stalk (peduncle); **female (pistillate) catkins are 1–2.5 cm long**; **yellow to light brown or black bracts** attach below each cottony seed.

HABITAT: Moist to wet meadows and marshes.

NATURAL HISTORY: The catkins of Geyer's willow appear with or just before the leaves, from mid-April through July. Geyer's willow is often found at higher elevations on both sides of the Cascades, along the Columbia River, but it is less common in the Willamette Valley. Willows attract ducks, marsh birds, songbirds, wildfowl and upland game-birds. Muskrats, porcupines and beavers eat the bark, and deer, rabbits and other small mammals eat the leaves and twigs. Willows are moderately palatable to livestock and big game. Willows also shade water, which lowers temperatures for fish, and they create habitat for many insects, which in turn provides food for many insectivorous wildlife species.

SIMILAR SPECIES: The slender shoots and twigs, short, narrow leaves and stubby catkins that give Geyer's willow its distinctive character also make it easy to distinguish from most other willows, but see Pacific willow (*S. lucida* ssp. *lasiandra*, p. 187).

NOTES: Early cultures have used willows to make medicines for many ailments, such as cuts, indigestion, worms and stomach problems.

NORTHWEST WILLOW
SOFT-LEAVED WILLOW • SANDBAR WILLOW
Salix sessilifolia

Salicaceae (Willow Family)
INDICATOR STATUS: FACW

GROWTH HABIT: Native shrub; **2–8 m tall**; **shrub or a small tree**; trunk sometimes to 10 cm thick; young twigs are usually thickly covered with downy hair; hairs fall off as stems age, but some fine hairs may remain.

LEAVES: Narrowly lance-shaped to oblong, 4–12 cm long, 1–3.5 cm wide; **tip pointed; base wedge-shaped**; densely hairy when young, becoming less so with age, but **fine spreading hairs remain on undersides**; scattered teeth around leaf edges; leaf stalks (petioles) are 1–5 mm long.

FLOWERS: In **catkins** that are **hairy** and have **yellowish cast due to many yellow bracts** among flowers; **female (pistillate) catkins are silvery green**, mostly 3–5 cm long (or up to 10 cm).

HABITAT: Willow swamps, streambanks and shorelines in valleys and floodplains.

NATURAL HISTORY: Northwest willow catkins appear with or after the leaves. Northwest willow occurs along the Columbia River and is common at lower elevations west of the Cascades as far south as the Umpqua Valley in Oregon. Like other willows, it provides forage for small game, shade for fish, food for songbirds and erosion control for shorelines and riverbanks.

SIMILAR SPECIES: Columbia River willow (*S. fluviatilis*, p. 190) is best distinguished from northwest willow by the **hairs on the undersides of the leaves**. Columbia River willow leaves have hairs that are pressed flat to the leaf undersides, while northwest willow leaves have spreading hairs (see illustrations on page 191).

A: *Northwest willow* (Salix sessilifolia).
B: *Male northwest willow catkins.*
C: *A female northwest willow catkin.*

COLUMBIA RIVER WILLOW • RIVER WILLOW
Salix fluviatilis

Salicaceae (Willow Family)
INDICATOR STATUS: OBL

GROWTH HABIT: Native shrub; **2–6 m tall**; **shrub or small tree**, trunk can get up to 10 cm thick; young stems are brown or green, are covered with many straight, closely pressed hairs and have milky-looking, waxy coating that mostly wears off as they grow; **stems become smooth and hairless, grayish brown and scaly with age**.

LEAVES: Alternate; **deciduous** (fall off by late November); relatively **long and narrow, linear to lance-shaped**, 5–15 cm long, 4–15 mm wide; young leaves are also covered with temporary, flattened hairs; **undersides of leaves tend to be waxy or whitish, with persistent, flattened hairs; leaf stalks (petioles) are short** (1–5 mm long); many leaves appear to be stalkless (**subsessile**); minute stipules at stalk bases soon wither and fall away.

FLOWERS: In catkins at ends of leafy branches; **female (pistillate) catkins are lax, silvery green**, mostly 4–10 cm long; **yellow bract** sits below each cottony seed on mature catkin.

HABITAT: Grows in shrub swamps, wet prairies and roadside ditches, and scattered on riverbanks and sandbars.

NATURAL HISTORY: Columbia River willow catkins appear in late summer after most of the leaves have developed. The **catkins remain on the plants into winter**. This is different from most other willows, in which the catkins usually appear in early spring and then fall away by mid-summer. Columbia River willow occurs at low to moderate elevations on both sides of the Cascades. It enhances wildlife habitat, helps control erosion and shades fish habitat. Deer and rabbits browse the stems and foliage of Columbia River willow, and small game-birds and songbirds eat the catkins and buds.

SIMILAR SPECIES: It can be very difficult to distinguish between Columbia River willow and northwest willow (*S. sessilifolia*, p. 189). One useful distinction can be found in the hairs on the undersides of the leaves. Columbia River willow leaves have hairs that are pressed flat to the leaf undersides, while northwest willow leaves have spreading hairs on the leaf undersides (see illustrations). Also, the leaves of Columbia River willow are usually narrower than those of northwest willow.

NOTES: *Salix*, the willow genus, was named from the ancient Celtic word *salis*, which means 'near the water.' Salicylic acid, the active ingredient in aspirin and similar pain relievers, was originally obtained from willows, and is named after the *Salix* genus.

Columbia River willow

northwest willow

A: *Columbia River willow* (Salix fluviatilis), *showing a female catkin.*
B, C: *Columbia River willow* (S. fluviatilis).
D: *Columbia River willow* (S. fluviatilis) *has flattened hairs on its leaf undersides, while northwest willow* (S. sessilifolia) *has spreading hairs.*

CHAPTER FIVE

WOODED WETLAND COMMUNITIES

Woodland communities, in the context of this book, are dominated by trees with either closed or open canopies, and they occur on stream and river bottoms and along the shores of lakes, ponds and sloughs. Shrubs are invariably a major component of the understory and are usually the dominant plants in gaps in the tree canopy.

Commonly called riparian or gallery forests, these communities are not always wetlands. In many cases, streams have cut a deep channel that floods only occasionally, and the riparian vegetation is composed almost entirely of upland species. Also, flooding streams often dump silt when water overflows the channel, which builds a natural levee along the stream that is slightly higher than the surrounding floodplain. Upland plants are more likely to grow close to the stream on the drier levees, while wetland plants often grow farther away on the floodplain. Winter rains often form shaded vernal pools in woodlands that last until May or June. Historically, seasonal flooding and beaver activity probably caused more extensive pooling than what occurs today.

Black cottonwood (Populus balsamifera *var.* trichocarpa) *and red-osier dogwood* (Cornus sericea) *commonly form communities along riverbanks in poorly drained soil.*

Slough sedge (Carex obnupta, *chapter 3) is a main component of the understory for many Oregon ash* (Fraxinus latifolia) *communities. This community type is now rare.*

As in shrub-dominated wetlands, the classification of communities in forested habitats is difficult. Along low stream and river banks, Oregon ash (*Fraxinus latifolia*) and red-osier dogwood (*Cornus sericea*) form a common community, often mixed with stands of Pacific willow (*Salix lucida* ssp. *lasiandra*, chapter 4). Black cottonwood (*Populus balsamifera* var. *trichocarpa*), red-osier dogwood and stinging nettle (*Urtica dioica*) often occur in a community adjacent to or just inland from the ash/dogwood community. This cottonwood/dogwood/nettle community is common

A young community of western red cedar (Thuja plicata) *and skunk cabbage* (Lysichiton americanum).

on stream terraces and often grows in linear formations associated with buried gravels where abundant water is available.

The most common and extensive woodland type on wet floodplains in the Willamette Valley is the ash bottom, which is dominated by Oregon ash. Two recently identified communities within ash bottoms that are characterized by their understory species are slough sedge/snowberry (*Carex obnupta/Symphoricarpos albus*) and short-scale sedge/stinging nettle (*Carex deweyana/Urtica dioica*). Seasonal pools provide valuable habitat for aquatic species, such as water crowfoot (*Ranunculus aquatilis*) and the rare howellia (*Howellia aquatilis*), both of which are described in chapter 1.

On higher ground, Oregon ash often intermixes with upland species, and large specimens of Pacific ninebark (*Physocarpus capitatus*), vine maple (*Acer circinatum*), hazelnut (*Corylus cornuta*) and oval-leafed viburnum (*Viburnum ellipticum*) dominate the shrub layer. Grand fir (*Abies grandis*) may also occur in many gallery forests, as can Pacific yew (*Taxus brevifolia*). Oregon white oak (*Quercus garryana*) often marks what was once the border between prairie, oak savanna and gallery forest.

In the northern Willamette Valley, and along tributary streams draining the Coast Range and Cascade foothills, western red cedar (*Thuja plicata*) and skunk cabbage (*Lysichiton americanum*) once formed a common community, similar to the ones at higher elevations. Many of these have been cleared for bottomland pasture, and red alder (*Alnus rubra*) and reed canary-grass (*Phalaris arundinacea*, chapter 3) now dominate.

The extent of bottomland forests in the Willamette Valley are now much diminished from what they were at the time of Euroamerican settlement. General Land Office survey notes from the 1850s show riparian forest to be as much as 8 km wide in some areas. Probably at least 40,000 hectares were cleared for agriculture and fuel wood, and most stands are now reduced to narrow strips along streams and rivers.

An old, mature community of western red cedar (Thuja plicata) *and skunk cabbage* (Lysichiton americanum).

Equisetaceae (Horsetail Family)
INDICATOR STATUS: FAC

GROWTH HABIT: Perennial native **horsetail**; **stems have regularly spaced nodes or joints** that can be pulled apart easily; **two stem types**: **sterile stems**, which are more common and last longer, and **fertile stems**, which produce spores and soon wither; **sterile stems are green**, 15–60 cm tall, 1.5–5 mm thick, with 10–12 minute ribs running lengthwise; dense whorls of branches (often mistaken for leaves) form at stem nodes; branches are 1–1.5 mm thick, sometimes branch again; **fertile stems appear before sterile stems, whitish to flesh colored** (become brownish just before withering), **unbranched**, 30 cm tall, 8 mm thick; **spore cone at tip**.

LEAVES: Reduced to **tiny scales** that are **fused into 6–14-toothed sheaths at stem nodes**.

FLOWERS: Horsetails **reproduce by spores**, and do not have flowers; **green spores** are produced in **flesh-colored cone at tip of fertile stem**.

HABITAT: Wooded wetlands, meadows, ditches, seeps and sandy or clay soils along lakes and streams.

NATURAL HISTORY: The fertile stems of common horsetail appear in early spring before the vegetative stems have grown tall enough to block spore dispersal by the wind. The spores have appendages on them that curl when wetted and uncurl when dried, which helps disperse the spores and move them deeper in the soil. Common horsetail is often found on disturbed sites.

SIMILAR SPECIES: Giant horsetail (*E. telmateia* ssp. *braunii*, FACW) is also found in wetlands in the Pacific Northwest. It is more robust and larger than common horsetail and the sheaths around its sterile stems have 14–18 teeth. Water horsetail (*E. fluviatile*, OBL) is a slender species that often forms extensive stands of identical individuals (clones) in standing water in marshes. It is often sparsely branched, and is distinguished by the very narrow, black scales on its stem. It also produces cones on its green stems. The difficulty in identifying horsetails is further complicated by the fact that common horsetail hybridizes with other species.

NOTES: Horsetails, also known as scouring rushes, have silica in their tissues, which makes them gritty. The silicated stems were used by Native Americans (and still used by some people today) to start 'hand-drilled' fires.

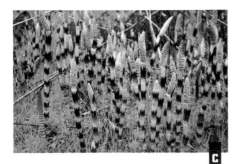

A: *Common horsetail* (Equisetum arvense). *The fertile stems are unbranched.*
B: *A sterile common horsetail* (E. arvense) *stem.*
C: *Giant horsetail* (E. telmateia *ssp.* braunii) *has more robust fertile stems.*

SKUNK CABBAGE
Lysichiton americanum

Araceae (Arum Family)
INDICATOR STATUS: OBL

GROWTH HABIT: Perennial native forb; flower stalks and **large, erect leaves** grow from underground stems.

LEAVES: Fleshy, ascending, lance-shaped to oblong-oval, **10–150 cm long**; leaf stalks (petioles) can be very short (giving impression that blades grow directly out of soil), or blades can be attached to long, fleshy stalks that grow up out of the muck.

FLOWERS: On upright, cylindrical, cone-like, fleshy spike (spadix) that looks like a corn cob; **spadix is enclosed in large, showy, yellow or cream-colored, sheathing bract (spathe)** that eventually falls away; **mature spadix is club-**like, 6–8 cm thick, up to 30 cm long; **flowers are tiny, yellowish green**.

HABITAT: Wooded wetlands, swamps, marshes and wet meadows.

NATURAL HISTORY: Skunk cabbage blooms in early spring before the leaves appear. The pungent odor of its flowers attracts flies and beetles as pollinators. It is found from sea level to high-elevation forests. Skunk cabbage remains fairly small when it grows in exposed areas, but when it is an understory plant in muddy, swamp-like conditions it can become massive. Skunk cabbage is often found in association with sedges (*Carex* spp.), small-fruited bulrush (*Scirpus microcarpus*) and lady fern (*Athyrium filix-femina*) in a variety of freshwater wetlands. During winter, the plant is not visible above ground, but in spring new growth quickly emerges from the old stem crowns (the central portion of the plant). The crown is the vegetative propagule, and often these can be salvaged from wet pasturelands or habitat that is designated for destruction. Skunk cabbage stems are eaten by muskrats and other animals, and bears apparently eat it in Alaska.

SIMILAR SPECIES: California false hellebore (*Veratrum californicum*, p. 197) is much taller than skunk cabbage and it has folded (pleated) leaves, many of which are attached all the way up the main stem. Skunk cabbage usually grows low to the ground and its leaves fan out from the base of the plant. Skunk cabbage leaves are also darker green and lack the prominent parallel veins found in California false hellebore leaves.

NOTES: Skunk cabbage can have the largest leaves of any species in the native flora of the Pacific Northwest. When bruised, the leaves exude a rank, pungent odor, which inspired the plant's common name. The flowers also stink. The entire plant contains calcium oxalate crystals, which can temporarily paralyze the salivary glands and cause the throat and tongue to swell and may constrict breathing. Ingestion is not recommended. Apparently, Native Americans used the underground stem to supplement their winter fare (presumably they cooked it). The peppery sap of the stems is also believed to have been used by earlier cultures to treat ringworm.

A: *Skunk cabbage* (Lysichiton americanum).
B: *Skunk cabbage in a typical habitat.*

Liliaceae (Lily Family)
INDICATOR STATUS: FACW+

GROWTH HABIT: Perennial native forb; **1–2 m tall**; conspicuously tall plant with lush new foliage (early in spring) that becomes tattered and discolored with age; stems are hairless (glabrous) near base, densely hairy above.

LEAVES: **Large, broad, oblong-lance-shaped**, 20–30 cm long, about half as wide; **coarse parallel veins** give rise to **conspicuous folds (pleats)** in blade; leaves look neat and perfect when they first emerge, but by flowering time they appear shredded and wilted.

FLOWERS: **Crowded on spreading, ascending branches of panicle**; flowers are **six-parted, small, dull white or slightly greenish, saucer- or bell-shaped**; tepals (undifferentiated petals and sepals) are **oval to lance-shaped**, 10–17 mm long.

HABITAT: Wet meadows and swampy ground associated with wooded wetlands.

NATURAL HISTORY: California false hellebore appears in early spring from lowlands to sub-alpine elevations. At higher elevations, it can be an indicator of the presence of elk, which use it for food and bedding in meadows.

SIMILAR SPECIES: See skunk cabbage (*Lysichiton americanum*, p. 196), which does not have pleated leaves or six-parted flowers.

NOTES: The young shoots, roots and thick underground stems (rhizomes) contain toxic alkaloids, and Native Americans used them medicinally in dilute doses.

A: *California false hellebore* (Veratrum californicum).

B: *California false hellebore leaves in early spring. The inevitable insect attacks and mammal browsers will shred and tatter the leaves by late spring.*

CANDY-FLOWER
SIBERIAN SPRING BEAUTY • MINER'S LETTUCE
Claytonia sibirica (Montia sibirica)

Portulacaceae (Purslane Family)

INDICATOR STATUS: FAC

Candy-flower (Claytonia sibirica).

GROWTH HABIT: Annual/perennial native forb; low-growing; **stems are 3–5 cm long, slender, weak and succulent**; occasionally produces short underground stems (rhizomes) and survives as short-lived perennial.

LEAVES: **Stem-base (basal) leaves are lance-shaped to oval**, 1–4 cm wide and as long or longer, **some are more linear** and are 3–10 mm wide; **leaf stalks (petioles) are long, slender**, 2–3 times longer than blades; pairs of **opposite, stalkless (sessile) leaves subtend flowers**, blades are **oval or heart-shaped**, 1–5 cm wide, up to 7 cm long, and have **pointed tips**.

FLOWERS: In **2–3 many-flowered racemes**; each dainty flower has **five notched petals** that are **completely pink or white with faint pink stripes**.

HABITAT: Ash swales, moist woods and wooded streambanks.

NATURAL HISTORY: Candy-flower blooms from middle to late May. The amount of sunlight a plant receives affects the color of the flowers. If candy-flower grows in deep shade, its flowers are deep pink, while plants growing in the sun have nearly all white flowers with fine pink veins.

SIMILAR SPECIES: Miner's lettuce (*Claytonia perfoliata*, formerly *Montia perfoliata*, FAC) has round upper leaves that completely surround the stem, and it grows in drier habitats than candy-flower. Also see water chickweed (*Montia fontana*), Howell's miner's lettuce (*M. howellii*) and narrow-leaf miner's lettuce (*M. linearis*) on page 104. The flowers of candy-flower look like those of Watson's willow-herb (*Epilobium ciliatum* ssp. *watsonii*, p. 175), but willow-herbs have very different leaves and their flowers have four petals, not five.

SMALL BEDSTRAW
Galium trifidum var. pacificum

Rubiaceae (Madder Family)
INDICATOR STATUS: FACW+

Small bedstraw (Galium trifidum *var.* pacificum).

GROWTH HABIT: Perennial native forb; **can be erect** or ascending, but it is **often lax and prostrate, scrambling** over other vegetation; **stems are slender, square**, unbranched on their lower portion but with several **branches from upper leaf axils**; **tiny, hooked barbs** point down along stem ridges.

LEAVES: In dainty **whorls** around stems; **4–6 leaves per whorl**; leaves are **linear, blunt**, 5–20 mm long, attached directly to stems (sessile); like stems, leaves have **tiny, hooked barbs**, which point down leaf along leaf edges.

FLOWERS: In groups of 1–3 on short stalks (peduncles); flowers are tiny, white.

FRUITS: Rounded, 1–1.75 mm thick.

HABITAT: Wooded areas beside streams and wooded ash swales.

NATURAL HISTORY: Small bedstraw blooms from June through July and sets seed quickly. Like its relatives, small bedstraw has brittle stems that are covered with tiny barbs. The barbs catch on passing animals and the weak stems break easily and are carried away, dispersing the seeds.

SIMILAR SPECIES: Generally, small bedstraw has fewer flowers than other bedstraws, of which there are many species in the valleys and floodplains of the Pacific Northwest. The most widespread is common bedstraw (*G. aparine*, FACU), also called cleavers, which has barbed hooks on its fruits. Common bedstraw and other upland bedstraw species grow in moist, shady places. See also wall bedstraw (*G. parisiense*, p. 130).

GREAT BETONY · HEDGE NETTLE
Stachys cooleyae

Lamiaceae, also called Labiatae (Mint Family)
INDICATOR STATUS: FACW

A

B

GROWTH HABIT: Perennial native forb; **60–150 cm tall**; stems are **simple, square, with bristles that curve down along stem ridges**.

LEAVES: Opposite; oval, 6–15 cm long, 2.5–8 cm wide; shallowly **notched along edges**; hairy (pubescent); leaf stalks (petioles) are 1.5–4.5 cm long.

FLOWERS: In series of **whorls in upper leaf axils; red to deep purple; covered with tiny, gland-tipped hairs; petals are small**, rather inconspicuous, fused together into **2-lipped tube**.

HABITAT: Common in swampy or marshy woods and along lakeshores and streambanks.

NATURAL HISTORY: Great betony blooms in July and August.

SIMILAR SPECIES: There are three smaller betonies that grow in wet meadows and prairies in the Pacific Northwest. Mexican betony (*S. mexicana*, FACW), which is chiefly a coastal resident but is found in the valleys from Eugene to Portland, has paler, generally pink or pink-purple flowers. Rigid betony (*S. rigida*, FACW-) does not extend as far north and is more likely to be found in the central and southern Willamette Valley. It has smaller leaves than great betony (3.5–9 cm long and 1–4 cm wide), and its upper leaves are generally stalkless while its lower leaves are often on stalks 1–4 cm long. Marsh betony (*S. palustris*, FACW+) is very similar to rigid betony, but its leaves are all stalkless, or some of the middle ones can be on very short stalks, well under 1 cm long. Great betony is easily confused with stinging nettle

C

(*Urtica dioica*, FAC+) when neither are in flower. Their stems and leaves are similar and both grow in moist habitats. However, the leaf stalks of great betony become smaller moving up the main stem, and in the top sections the leaves may appear to be stalkless, while stinging nettle leaves always have a noticeable stalk. Stinging nettle leaves also have more sharply pointed tips. It is easy to distinguish these two species if they are in flower. Stinging nettle has long, stringy beads of tiny, greenish or brown flowers, while great betony has distinct spikes of purple flowers. Also, stinging nettle does not have a distinctive odor, and if you touch it you will soon discover its abundant, stinging hairs, resulting in a positive identification.

NOTES: Great betony has a strong, disagreeable odor that can sometimes be detected before sighting the plant.

D

A: *Great betony* (Stachys cooleyae).
B: *Stinging nettle* (Urtica dioica).
C: *Marsh betony* (S. palustris).
D: *Rigid betony* (S. rigida).

WESTERN WATER-HEMLOCK
Cicuta douglasii

Apiaceae, also called Umbelliferae (Carrot Family)
INDICATOR STATUS: OBL

GROWTH HABIT: Perennial native forb; **0.5–2 m tall; stems are hollow and stout**, grow from **swollen base** that also produces **fleshy, chambered rootstock**; lower stems sometimes root from swollen nodes.

LEAVES: 1–3 times pinnately compound; leaflets towards leaf tip are simple while those towards base are further divided one or two times; leaflets are coarse, thick, **lance-shaped**, 3–10 cm long, 6–35 mm wide, with **serrated edges**; short sheath encloses stem just below point of leaf attachment.

FLOWERS: In several **compound umbels**; umbellet stalks (rays) are 2–6 cm long; many **tiny, white flowers** per umbellet.

HABITAT: Grows in wet meadows and pastures, and along or in streams and ditches.

NATURAL HISTORY: Western water-hemlock occurs from June through August. Its seeds are eaten by mallards and other waterfowl.

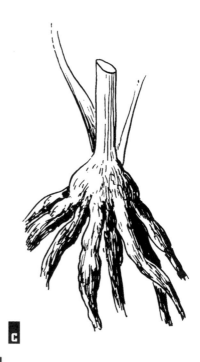

SIMILAR SPECIES: Poison-hemlock (*Conium maculatum*, FAC+), which is often confused with western water-hemlock, has purple-spotted stems and more highly divided leaves and grows in generally drier habitats, though often this is along ditches near the water. Water parsnip (*Sium suave*, OBL) leaves have a similar arrangement to those of western water-hemlock, but the leaflets are much narrower in water parsnip. Western water-hemlock can also be distinguished from water parsnip and other similar species by the **unique pattern of veins** on its leaves. The **prominent side veins end at the notches** along the leaflet edge, rather than at the tips of the serrations, as in water parsnip (see drawings and photo). See also kneeling angelica (*Angelica genuflexa*, p. 204) and water parsley (*Oenanthe sarmentosa*, p. 66).

NOTES: All parts of western water-hemlock are **poisonous** to people and livestock. The swollen, fleshy roots have an unpleasant odor and contain the highest concentration of poison.

A: *Western water-hemlock* (Cicuta douglasii).

B: *Detail of the rooting node on a western water-hemlock stem.*

C: *A western water-hemlock rootstock.*

D: *The chambered rootstocks of western water-hemlock help identify this very poisonous plant.*

E: *Western water-hemlock leaves are one to three times pinnately divided.*

KNEELING ANGELICA
Angelica genuflexa

Apiaceae, also called Umbelliferae (Carrot Family)
INDICATOR STATUS: FACW

GROWTH HABIT: Perennial native forb; **often 1–2 m tall**; one leafy main stem per plant. **LEAVES:** Compound; **triangular** in outline, **or narrower at base and tip and wider in middle**, 10–80 cm wide; **twice divided** into **three triangular to lance-shaped leaflets,** mostly 4–10 cm long and 1.5–5 cm wide; some leaflets are deeply 3-lobed or again divided; leaflet **edges are noticeably saw-toothed (serrate)**; main leaf stalk (petiole) is conspicuously **bent backwards (reflexed) where each pair of leaflets attaches**, which gives the leaf a 'genuflected' or 'kneeling' character.
FLOWERS: In several **compound umbels**; **25–50 umbellets** per umbel, each subtended by **ring of leaf-like bractlets; umbellet stalks (rays) are 2–7 cm long**; several, **tiny, white or pinkish flowers**, each on slender stalk (pedicel).
FRUITS: Round to egg-shaped, 3–4 mm in diameter when mature.
HABITAT: Moist woods and marshes from valleys to higher elevations.
NATURAL HISTORY: In most places in the Pacific Northwest, kneeling angelica can be found in bloom during June and July. It is associated with forested wetlands and sometimes shrub swamps.
SIMILAR SPECIES: Other angelicas, especially sharp-tooth angelica (*A. arguta*, FACW), can be confused with kneeling angelica, but they lack the **ring of bractlets below the umbellets**. Also, kneeling angelica is the only species with 'genuflected' leaves. Western water-hemlock (*Cicuta douglasii*, p. 202) has narrower leaves and much narrower (3 cm wide) leaf sheaths, and its underground stems (rhizomes) are divided into distinct chambers. Poison-hemlock (*Conium maculatum*, p. 202) has purple-spotted stems, much more finely divided leaves and a mouse-like odor, and grows in drier habitats. Water parsley (*Oenanthe sarmentosa*, p. 66) has similar umbels of tiny white flowers, but each umbel has a pair (not a ring) of leaf-like bracts. Also, the umbels of water parsley are smaller and their rays are fewer (10–20) and shorter (1.5–3 cm long). The leaves of water parsley are much more divided than those of kneeling angelica, and resemble culinary parsley leaves.
NOTES: This stout angelica is much larger than the other angelicas found in wetlands. Kneeling angelica is related to poison-hemlock and water-hemlock, so it is possibly **poisonous** and should not be eaten.

A: *Kneeling angelica* (Angelica genuflexa).
B: *The bent leaf stalks of kneeling angelica are distinctive.*

TALL MANNAGRASS
Glyceria elata

Poaceae, also called Gramineae (Grass Family)

INDICATOR STATUS: FACW+

GROWTH HABIT: Perennial native grass; 1–1.5 m tall; grows singly (especially in wooded wetlands) or in small colonies along streambanks; stems are erect, **curving somewhat at bases**.

LEAVES: Narrow, soft, flat, ribbon-like, 6–10 mm wide; leaf **sheaths are closed nearly their full length**; leaf blades and sheaths feel rough when rubbed from tip to base; **ligules** at top of sheath are **3–6 mm long**.

FLOWERS: In **open, spreading panicles** that tend to **droop, are 15–25 cm long** and have a **purplish cast** to them; lemmas have **prominent, parallel veins**.

HABITAT: Shaded Oregon ash swales, swamps and wooded streambanks.

NATURAL HISTORY: Tall mannagrass occurs from May through August. It provides food and cover for muskrats and birds, especially waterfowl, such as gadwalls, mallards and wood ducks, and it is also browsed by deer. Tall mannagrass will bloom even when it is entirely submerged.

SIMILAR SPECIES: There are several aquatic or semi-aquatic mannagrasses. Northern mannagrass (*G. borealis*, OBL), also called floating mannagrass, is the most aquatic, and it grows in 30 cm or more of water. Its leaves are narrower than those of the other mannagrasses (2–7 mm wide, as compared to 4–13 mm wide), and they are usually flattened and often bowed at the base (decumbent). Western mannagrass (*G. occidentalis*, p. 80) prefers shallower water and is less common than northern mannagrass. Both northern mannagrass and western mannagrass have narrower, more confined panicles than the other mannagrasses, and they have longer spikelets that are pressed flat to the stem. Tall mannagrass and reed mannagrass (*G. grandis*, formerly *G. maxima*, OBL) both grow in more semi-aquatic or wooded habitats and have erect stems and large, spreading, open panicles with short spikelets. They are hard to distinguish from one another, and the only reliable distinction between them is in a part of the floret called the palea. In tall mannagrass the **tip of the palea is narrowly notched**, while in reed mannagrass it

is jagged or widely notched. A fairly reliable distinction can also be made by pulling a leaf blade between your thumb and forefinger from the tip to the base. In tall mannagrass this produces a **rough, sandpapery feel**, a condition that is technically described as **retrorsely roughened**, while in reed mannagrass the leaf feels much more smooth. Fowl mannagrass (*G. striata*, OBL), which is not aquatic, is generally smaller than the other mannagrasses and grows at higher elevations (generally above 1,500 m). Like tall mannagrass and reed mannagrass, it has open panicles of short spikelets, but its panicles are narrower. Canadian bluegrass (*Poa compressa*, FACU) can also be mistaken for a mannagrass, but it has open leaf sheaths, while mannagrasses have closed sheaths. The panicles of Canadian bluegrass are narrow like those of northern mannagrass and western mannagrass, but its spikelets are much shorter.

NOTES: The genus name, *Glyceria*, means 'sweet,' and both it and the common name 'mannagrass,' refer to the sweetness and palatability of the grain to livestock.

A, B, C: *Tall mannagrass* (Glyceria elata). **C**

Cyperaceae (Sedge Family)
INDICATOR STATUS: FACU

GROWTH HABIT: Perennial native grass-like; **20–100 cm tall**; grows in loose clumps or densely packed clusters; stems are quite **leafy, tend to fall over and spread out** along ground.

LEAVES: On **lower parts** of stems; **flat**, usually **2–5 mm wide**.

FLOWERS: In **loosely compact clusters** of **4–10 green or greenish-tan spikes** at ends of long, sometimes trailing stalks (peduncles); spikes are stalkless (**sessile**), 7–20 mm long, more closely grouped towards top of cluster, **lowest spike can be quite remote**; both male and female flowers in each spike, with **female flowers occurring above male flowers**; short, narrow, **leaf-like bracts** attach directly below spikes; these **bracts have inflated sheaths**.

HABITAT: Wooded wetlands and stream edges in at least partial shade.

NATURAL HISTORY: Short-scale sedge occurs from sea level to the treeline of alpine ecosystems.

SIMILAR SPECIES: Henderson's sedge (*C. hendersonii*, FAC) can easily be mistaken for short-scaled sedge. The most noticeable difference between the two is in the size and arrangement of their leaves. The leaves of Henderson's sedge are wider (5–14 mm) and arise from the base of the stem, while the leaves of short-scale sedge arise from the lower parts of the stems.

A: *The typical appearance of short-scale sedge* (Carex deweyana) *in a wooded wetland.*
B: *Short-scale sedge seed heads.*
C: *Short-scale sedge.*

WOOLLY SEDGE
Carex lanuginosa (C. pellita)

Cyperaceae (Sedge Family)
INDICATOR STATUS: OBL

GROWTH HABIT: Perennial native grass-like; 30–100 cm tall; grows singly or as many plants clustered together, which arise from creeping underground stems (rhizomes).

LEAVES: Few well-developed leaves, borne well above stem base, **long, fairly flat, 2–5 mm wide**; fibrous **sheaths** surround stem bases; many **shredded, reddish scales** (reduced leaves) occur at stem bases.

FLOWERS: In groups of **straw-colored or yellowish-brown, cylindrical spikes**; **upper 1–3 spikes contain male flowers**; uppermost male spike is 2–5 cm long, sometimes falls away; **lower 2–3 spikes contain female flowers**; female spikes are 2–3 cm long, **stalkless (sessile) except lowest spike**, which may have short, slender, erect stalk (peduncle); perigynia are 3.3–5 mm wide, covered with **fine, silky to velvety, spreading hairs** (diagnostic for this species); **leaf-like bract** at base of each spike; usually only **lowest bract is longer than spikes**.

HABITAT: Wooded wetlands; occasionally in ditches, wet meadows and prairies.

NATURAL HISTORY: Woolly sedge is often found in shallow, standing water in the valleys and floodplains of the Pacific Northwest. It is also common in moderate-elevation wetlands west of the Cascades and occurs over much of the United States.

SIMILAR SPECIES: See Columbia sedge (*C. aperta*, p. 82), slough sedge (*C. obnupta*, p. 160), and inflated sedge (*C. vesicaria* var. *major*, p. 83). Woolly sedge is the only one of these three with **velvety hairs** on its perigynia.

NOTES: Woolly sedge gets its common name from the minute, velvety hairs on its perigynia, a feature that is hard to see without a hand lens.

A, B, C: *Woolly sedge (Carex lanuginosa).*

Cornaceae (Dogwood Family)
INDICATOR STATUS: FACW

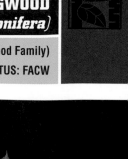

Red-osier dogwood (Cornus sericea).

GROWTH HABIT: Native **shrub**; **2–6 m tall**; freely branching and multi-stemmed, in dense thickets in wetlands; branches, especially new ones, are bright, **glossy red to reddish purple**; lower, **older stems are grayish brown**; twigs and branches have **crescent-shaped leaf scars**.

LEAVES: Opposite; **elliptical to egg-shaped (ovate)**, 4–12 cm long, about half as wide; leaf tips are abruptly **pointed**; leaf undersides are somewhat pale and hairy; upper surfaces are green and have **prominent veins** that converge towards leaf tip; leaves turn various vibrant shades of orange and red before dropping in fall; if leaf is pulled apart carefully, veins remain connected to midrib by thin silvery strands (vascular bundles).

FLOWERS: In **flat-topped clusters** from leaf axils at branch tips; flowers are **small, white**.

FRUITS: White or porcelain-blue, berry-like drupes; about 9 mm wide; central stone or pit.

HABITAT: Wooded and shrub swamps, especially along creeks and rivers.

NATURAL HISTORY: Red-osier dogwood blooms from May through June. It is commonly associated with willows (*Salix* spp.) and black cottonwood (*Populus balsamifera* var. *trichocarpa*) in riparian systems. The fruits provide good food for songbirds, especially cedar waxwings, and deer and other browsers eat the buds.

SIMILAR SPECIES: Nuttall's dogwood (*C. nuttallii*, NOL) is distinguished from red-osier dogwood by its young twigs, which are not red. Indian plum (*Oemleria cerasiformis*, FACU) has alternate leaves that are narrower and less leathery than those of red-osier dogwood.

NOTES: The name 'dogwood' derives from the Scandinavian *dag*, meaning 'skewer,' because of early uses of the stems for stretching, drying and roasting meat. 'Osier' is a Middle English word for the tough, pliant willow branches used in basketry. It derives from the French *osière*, which is linked to a Gaulish word that means 'river-bed.'

SNOWBERRY
Symphoricarpos albus

Caprifoliaceae (Honeysuckle Family)
INDICATOR STATUS: FACU

GROWTH HABIT: Native **shrub**; **0.5–2.5 m tall**; can form dense thickets from underground shoots (rhizomes); stems are highly branched.

LEAVES: Opposite; **simple**; usually **small**, 1.5–5 cm long, 1–3.5 cm wide; **narrowly oval**; bright green above; paler green below, sometimes with fine, sparse hairs; **smooth-edged (entire), toothed or irregularly lobed**; leaves along new shoots can be irregular in shape and appear lobed, like oak leaves.

FLOWERS: Clustered at ends of branches; flowers are **small, pink or white**.

FRUITS: Round, puffy, white, berry-like **drupes**, up to 1–1.5 cm in diameter; fruit pops when squeezed; **poisonous** to humans.

HABITAT: Wooded wetlands; becomes especially thick around wetland edges, creating thickets.

NATURAL HISTORY: Snowberry blooms in spring from mid-May to July. It is a common shrub in both upland forests and in the understory of wooded wetlands. It is also associated with Nootka rose (*Rosa nutkana*), clustered rose (*R. pisocarpa*), sweetbrier (*R. eglanteria*) and Douglas' spiraea (*Spiraea douglasii*) in shrub-swamp thickets at the edges of lakes, or invading the edges of wet prairies not recently burned. The foliage of snowberry is fragrant when wet. Lingering snowberry fruits provide food for wildlife in fall and winter when other sources are limited.

SIMILAR SPECIES: Creeping snowberry (*S. mollis*, NOL) is a trailing vine that grows in wetlands and upland forested areas. Honeysuckle species, especially twinberry (*Lonicera involucrata*, FAC), can resemble snowberry, but twinberry has prominent, paired flowers and inky, blue-black fruits. Also, twinberry leaves are at least twice as large as those of snowberry. Ocean spray (*Holodiscus discolor*, NOL) is another shrub found in moist forested areas, but it has 'foamy' cascades of white flowers.

A, B: *Snowberry* (Symphoricarpos albus).
C: *Snowberry leaves on vigorous new shoots can have irregular shapes.*

Rosaceae (Rose Family)
INDICATOR STATUS: FACW-

GROWTH HABIT: Native **shrub**; **2–4 m tall**; **deciduous**; branches are long, arching or erect, are **covered with star-shaped hairs**, and continually shed **peeling layers of papery bark**.

LEAVES: Palmately 3–5-lobed, with **doubly saw-toothed (serrate) edges** (like small versions of maple leaves); 4–8 cm long, nearly as wide; glossy, dark green above; undersides are paler because of the star-shaped hairs (visible with a 10x hand lens).

FLOWERS: In **rounded, head-like clusters**; flowers are **small, white**, with **bright pink stamens**; **petals are nearly round**, about 4 mm long.

FRUITS: Dry, many-seeded follicle.

HABITAT: Ash swales, wooded streambanks and lakeshores.

NATURAL HISTORY: Pacific ninebark blooms from May to June. It occurs along the west side of the Cascades from southern Alaska to northern California.

SIMILAR SPECIES: Oval-leaf viburnum (*Viburnum ellipticum*, NOL), ocean spray (*Holodiscus discolor*, NOL) and goatsbeard (*Aruncus sylvester*, formerly *A. dioicus*, NOL) are three other white-flowered shrubs that can be mistaken for Pacific ninebark. Pacific ninebark is best distinguished from them by its leaves and peeling bark. Oval-leaf viburnum is 1–3 m tall, has its flowers in rounded, head-like clusters, and has simple, broadly elliptical leaves that are 3–8 cm long. Goatsbeard is 1–2 m tall and its flowers are in erect, spike-like clusters along the branches of a panicle. Its leaves are up to 15 cm long and 8 cm wide overall, and are divided (sometimes twice) into three leaflets. Ocean spray is 1–3 m tall, with flowers in diffuse, lax panicles that are 10–17 cm long, and egg-shaped (ovate) leaves that are 4–7 cm long, with 15–25 shallow lobes around the edges.

NOTES: The species name, *capitatus*, from the Latin for 'a head,' describes the head-like flower clusters, which are rounded above but flattened at the base.

A: *Pacific ninebark* (Physocarpus capitatus).
B: *The bark of Pacific ninebark sheds in peeling layers.*

SALMONBERRY
Rubus spectabilis

Rosaceae (Rose Family)
INDICATOR STATUS: FAC+

Salmonberry (Rubus spectabilis). *The orange or sometimes red fruits look like raspberries.*

GROWTH HABIT: Native **shrub**; **1–3 m tall**; sometimes in dense thickets from underground stems (rhizomes); **woody stems (canes) erect to arching,** usually **prickly** (less so upwards).

LEAVES: **Pinnately compound**; **3 or 5 oval leaflets, taper to pointed tips, toothed around edges;** end leaflet largest (4–5 cm long).

FLOWERS: Borne singly or in pairs; **petals are hot pink to red, oval,** 13–22 mm long.

FRUITS: **Raspberry-like clusters** of **orange, yellow or purplish-red** drupelets.

HABITAT: Seeps and swales in understory of moist woods, wet slopes and wooded streambanks.

NATURAL HISTORY: Salmonberry blooms from April to June, and its fruits ripen in late July. These fruits can be an important food source to grouse, pheasants, robins, orioles, thrushes, tanagers and towhees, racoons, chipmunks and squirrels. Salmonberry leaves are sometimes browsed by deer and rabbits.

SIMILAR SPECIES: Salmonberry can be confused with other *Rubus* species, especially thimbleberry (*R. parviflorus*, FACU+). However, thimbleberry is thornless and its leaves are simple (not compound), velvety, broad and lobed like a maple leaf. Ocean spray (*Holodiscus discolor*, NOL) and goatsbeard (*Aruncus sylvester*, formerly *A. dioicus*, NOL) are easily distinguished from salmonberry by their thornless stems and 'foamy' cascades of white flowers. All of these look-a-likes belong to the rose family.

NOTES: Salmonberries are edible and tasty when ripe. They are generally sweeter when orange and more sour when red. Native Americans ate them and David Douglas reported that they also ate the acidic young shoots with dried salmon roe to cut the brine, which may be the original reason for the common name. It is more widely held, however, that the common name reflects the color of the berries, or that the drupelets look like salmon eggs.

OREGON ASH
Fraxinus latifolia

Oleaceae (Olive Family)
INDICATOR STATUS: FACW

GROWTH HABIT: Native tree; **10–20 m tall**; **deciduous**; rare old-growth specimens invariably have hollow trunks that can reach 2 m in diameter; bark on the **trunk is rough, grayish brown**, becomes **deeply ridged with age**; ridges form **diamond or diagonal pattern**.

LEAVES: **Pinnately compound**, usually with **five or seven leaflets**; leaflets are **oblong to oblong-ovate**, up to 15 cm long; **light green above, paler below**.

FLOWERS: In crowded panicles; **tiny** (3 mm across), inconspicuous; **sexes are on separate trees**; male flowers are yellowish; female flowers are greenish.

FRUITS: Long, single-winged **samaras**, about 3–5 cm long; hang together **in dense clusters**.

HABITAT: Wooded wetlands, lakeshores and edges of low-elevation streams west of Cascades.

NATURAL HISTORY: Oregon ash blooms in spring, and its fruits are generally produced in June and remain into mid-summer. Oregon ash woodlands, often called ash swales and gallery forests, are found throughout the interior valleys of the Pacific Northwest. Willows (*Salix* spp.), hawthorns (*Crataegus* spp.), alders (*Alnus* spp.), red-osier dogwood (*Cornus sericea*), snowberry (*Symphoricarpos albus*), Pacific ninebark (*Physocarpus capitatus*) and slough sedge (*Carex obnupta*) are often found in the shrubby understory and ground cover of an ash swale. Oregon ash fruits are of moderate importance to seed-eating birds and mammals. The bark and limbs of Oregon ash are often encrusted or draped with lichens, which gives the trees a frosty appearance.

SIMILAR SPECIES: Red alder (*Alnus rubra*, p. 214), black cottonwood (*Populus balsamifera* var. *trichocarpa*, p. 215) and some willows (pp. 184–91) can resemble Oregon ash, but Oregon ash is distinguished by its compound leaves. Red alder, black cottonwood and willows all have simple, undivided leaves. Oregon ash leaflets that have detached from the main leaf stem and fallen to the ground may look like willow leaves.

A: *Oregon ash (Fraxinus latifolia).*
B: *Oregon ash, showing the samaras.*
C: *The bark of Oregon ash has a distinctive diamond pattern.*

RED ALDER
Alnus rubra

Betulaceae (Birch Family)
INDICATOR STATUS: FAC

GROWTH HABIT: Native tree; **to 25 m tall**; **deciduous**; trunk is up to 80 cm thick; **bark is smooth, mottled gray and white**.

LEAVES: Shape varies from **oblong to oval to elliptical-oval**, usually **broadest in middle** of blade; 5–15 cm long; **upper surface is dull, dark green**, smooth; **underside is dull lime green to gray**, and slightly **rust colored in older leaves**; leaf **edges roll inwards** slightly on underside and are **irregularly single-toothed**.

FLOWERS: Male flowers are arranged on long, narrow, cylindrical, jointed, flexible, drooping catkins; female flowers and mature seeds are borne on **small, clustered, stubby 'cones'** that are 1.5–2 cm long.

HABITAT: Wooded wetlands, especially along streams, rivers, gravel bars and new, wet clearings; occasionally in shrub swamps.

NATURAL HISTORY: Red alder catkins appear in early spring before the leaves emerge. Red alder is a pioneer on recent alluvial deposits, creating alder flats. It is a dominant of disturbed sites, especially on coastal hills. Red alder buds and seeds may be eaten by upland game-birds, wildfowl and songbirds, and the twigs and foliage are eaten by beavers, rabbits, deer and other browsers. Red alder grows quickly and rapidly, and because it fixes nitrogen it improves soil fertility, especially in clear cuts or burned areas.

SIMILAR SPECIES: Red alder can be mistaken for other shrubby or tree-like species when it is young. It is most easily distinguished from other alder species, and any other shrub look-alikes, by the single row of teeth along its inrolled leaf edges. Sitka alder (*A. sinuata*, FACW) leaves are distinctly double-toothed, and its leaf edges are not at all inrolled.

A: *Red alder* (Alnus rubra). *The female flowers are bone in stubby 'cones.'*
B: *The bark of a red alder that is approximately 70 years old.*

BLACK COTTONWOOD • BALSAM POPLAR
Populus balsamifera var. *trichocarpa* (*P. balsamifera*)

Salicaceae (Willow Family)
INDICATOR STATUS: FAC

GROWTH HABIT: Native tree; **up to 60 m tall**; deciduous; trunk is usually less than 1 m in diameter, but can reach 2 m; bark is light colored, rough; winter buds are sticky and sweet-smelling.

LEAVES: Leaf shape is variable, can be **heart-shaped or triangular** (usually broadest below middle), with a **pointed tip**, 5–15 cm long, 3–9 cm wide; **shiny, dark green above; pale green to white below**; leaf edges are **shallowly toothed**; leaf stalks (petioles) are round and one to two thirds the length of the leaves.

FLOWERS: In **long, sticky catkins** that hang below leaves and stems towards ends of limbs; **male (staminate) catkins are 2–3 cm long**, soon fall off; **female (pistillate) catkins are 8–20 cm long, cottony.**

FRUITS: Round, green, hairy capsules; seeds are surrounded by fluffy, white hairs (look like 'seedy' cotton balls when they disperse in wind).

HABITAT: Wet sites, lakeshores, stream and river banks, sloughs and swamps, as well as old or inactive river beds.

NATURAL HISTORY: Black cottonwood blooms from late April through June. It is found in flood-plain forests and along riverbanks throughout the Pacific Northwest, often with alders (*Alnus* spp.) and willows (*Salix* spp.). Black cottonwood is usually sweetly fragrant, especially when the leaves are emerging from their buds.

SIMILAR SPECIES: Quaking aspen (*P. tremuloides*) is a smaller tree (up to 25 m tall) with smooth bark. Although both black cottonwood and quaking aspen have generally heart-shaped leaves, quaking aspen has much smaller, compact leaves (up to 7 cm long) with flat leaf stalks, causing them to flutter in the slightest breeze. Black cottonwood can be distinguished from the willows (pp. 184–91) by its large, **heart-shaped leaves** and by the way its fruits are attached along the **long, draping catkins** like a necklace of beads. The leaves of alders (p. 214) are similar in shape to those of black cottonwood, but alder leaves have prominent veins that give them a pleated look, unlike the smooth leaves of black cottonwood.

A

B

A, B: *Black cottonwood* (Populus balsamifera var. trichocarpa).

WESTERN RED CEDAR
PACIFIC RED CEDAR • GIANT CEDAR
Thuja plicata

Cupressaceae (Cypress Family)

Indicator Status: FAC

A

GROWTH HABIT: Native tree; **up to 70 m tall**; grows from one **well-developed trunk** that often forms **buttresses** (an adaptation to moist or boggy soils); branches are **spreading to drooping, upturned** at tips; bark is **gray to reddish, soft, thin**, constantly **shedding in long, fibrous strips**.

B

LEAVES: Small, **yellowish green, scale-like**; occurring in opposite, **horizontally flattened pairs** that **overlap** one another; **evergreen**, but die after 3–4 years and sometimes turn orange-brown before they fall off; new leaves are perpetually regenerated as old leaves are shed.

FLOWERS: The reproductive structures are in cones (not flowers); **male cones are narrowly cylindrical**, about 2 mm long, flexible, dangling, appearing segmented; **female cones are stubby**, 8–10 mm long, occurring **in clusters**; each female cone consists of three **opposite pairs of seed-bearing scales** and one narrow, sterile pair at tip; **seeds are 4–7 mm long**.

HABITAT: Commonly in moist to wet soils, usually in forests on seepage and alluvial sites; also in drier uplands.

NATURAL HISTORY: Cedars are members of the gymnosperm class of seed-bearing plants, which includes pines and firs, among others, and as such their seeds are 'naked,' and not contained within an ovary. Western red cedar pro-

C

D

duces pollen from April through May. It produces nesting habitat and protective cover for many songbirds, and cedar waxwings eat its immature, blue-black cones. Cedars grow slowly and live for hundreds of years.

SIMILAR SPECIES: Western red cedar is the largest cedar in our region and it has relatively large, yellowish-green leaves. Port Orford cedar (*Chamaecyparis lawsoniana*, FACU+) has a pleasant aroma and yellowish, flattened leaves with long, clasping bases. Incense cedar is generally smaller than the other cedars in our region. Alaska cedar (*Chamaecyparis nootkatensis*, FAC) has tiny, blue-green leaves. Incense cedar (*Calocedrus decurrens*, NOL) has distinctively furrowed, felty bark and small, blue-green leaves. Port Orford cedar is restricted to southwestern Oregon, from Winchester Bay, south to the Klamath Mountains, where it occurs in serpentine fens, riparian areas and in uplands.

NOTES: Western red cedar often has a 'candelabra' of multiple tops when it is old or if it is growing in wet areas. Western red cedar was very important to the Native Americans of the Pacific Northwest, who used its bark to make clothes, soft linings, blankets, ropes, baskets and houses. Western red cedar's soft wood was used to make totem poles and canoes, and other parts were used for medicine and charms.

E

A: *Western red cedar* (Thuja plicata).
B: *Western red cedar* (T. plicata) *bark.*
C: *The bark of this old western red cedar was peeled away for use by Native Americans.*
D: *Port Orford cedar* (Chamaecyparis lawsoniana).
E: *Incense cedar* (Calocedrus decurrens), *with a detail of the small, overlapping leaves.*
F: *Western red cedar* (T. plicata), *showing the female cones.*

F

APPENDIX 1

Species that occur in more than one habitat.

The plants described in this book are assigned to chapters based on the habitat in which they are most commonly found. However, many plant species can occur in more than one habitat—as indicated by the secondary habitat icons in the header bars—so a plant you are trying to identify may be described in a chapter other than the one for the habitat in which you found it.

The following chart should help reduce the time needed to identify plants that occur in more than one habitat. For each habitat, the chart lists the plant species that sometimes occur in that habitat but are described elsewhere in the book. The plant species have been subdivided into four growth forms—non-woody plants, grasses and grass-likes, shrubs and trees—and each species is followed by its page reference.

Submerged and Floating

Non-woody plants
creeping spearwort 59
water-starworts 60
false loosestrife 62
American brooklime 64
narrow-leaved water-plantain 71
American water-plantain 72
water smartweed 73
yellow marshcress 76

Grasses and grass-likes
western mannagrass 80
rice cut-grass 81
inflated sedge 83
needle spike-rush 85
ovate spike-rush 86
small-fruited bulrush 87
soft-stem bulrush 88
pointed rush 90
dagger-leaf rush 92

Marshy Shore

Non-woody plants
Mexican water fern 31
purple-fringed riccia 32
common duckweed 34
tapegrass 38
marsh pennywort 39
watercress 40
howellia 48
coontail 51
common bladderwort 54
water crowfoot 55
Nuttall's quillwort 97
western yellowcress 107
water-plantain buttercup 109
bog trefoil 115
small-flowered forget-me-not 120
fragrant popcornflower 121
wild mint 123
pennyroyal 124
bog St. John's-wort 128
moneywort 129
leafy beggarticks 139
cow parsnip 176
skunk cabbage 196

Grasses and grass-likes
large barnyard grass 142
reed canary-grass 143
American slough grass 153
slough sedge 160
creeping spike-rush 162
toad rush 163
taper-tipped rush 164
jointed rush 166
soft rush 170

Trees
black cottonwood 215

Wetland Prairie

Non-woody plants
watercress 40
false loosestrife 62
American brooklime 64
skullcap speedwell 65
water parsley 66
cattail 68
narrow-leaved water-plantain 71
American water-plantain 72
waterpepper 74
purple loosestrife 75
yellow marshcress 76
seaside trefoil 77
nodding beggarticks 78
Watson's willow-herb 175
common horsetail 195
candy-flower 198
small bedstraw 199
great betony 200

Grasses and grass-likes
western mannagrass 80
rice cut-grass 81
Columbia sedge 82
awl-fruited sedge 84
needle spike-rush 85
ovate spike-rush 86
small-fruited bulrush 87
hard-stem bulrush 88
pointed rush 90
Coville's rush 91
dagger-leaf rush 92
woolly sedge 208

Shrubs
serviceberry 177
Douglas' hawthorn 178
Nootka rose 182
northwest willow 189
Columbia River willow 190

Trees
Oregon ash 213

Shrub Swamp

Non-woody plants
water parsley 66
nodding beggarticks 78
large-leaf avens 113
northwest cinquefoil 114
seep-spring monkeyflower 127
wall bedstraw 130
wetland asters 134
California false hellebore 197
small bedstraw 199

Grasses and grass-likes
reed canary-grass 143
slender rush 169

Shrubs
red-osier dogwood 209
snowberry 210
Pacific ninebark 211

Wooded Wetland

Non-woody plants
water parsley 66
seaside trefoil 77
curly dock 102
large-leaf lupine 116
cut-leaf water-horehound 122
wetland asters 134
Watson's willow-herb 175
cow parsnip 176

Grasses and grass-likes
small-fruited bulrush 87
hard-stem bulrush 88
Kentucky bluegrass 152
slough sedge 160
slender rush 169
spreading rush 171

Shrubs
serviceberry 177
Douglas' spiraea 181
Nootka rose 182
Scouler's willow 184

APPENDIX 2

Indicator status assignments.

Abies grandis . . . . FACU
Acer circinatum . . . . FACU
Agrostis
  alba
    var. alba . . . . FAC
    var. palustris . . . . FAC+
  capillaris . . . . FAC
  diegoensis . . . . NOL
  idahoensis . . . . FACW
  stolonifera . . . . FAC
Alisma
  gramineum . . . . OBL
  lanceolatum . . . . OBL
  plantago-aquatica . . . . OBL
Allium amplectens . . . . NOL
Alnus
  rubra . . . . FAC
  sinuata . . . . FACW
Alopecurus
  aequalis . . . . OBL
  geniculatus . . . . FACW+
  pratensis . . . . FACW
Amelanchier alnifolia . . . . . . FACU
Anabaena sp. . . . . NOL
Anaphalis margaritacea . . . . . NOL
Angelica
  arguta . . . . FACW
  genuflexa . . . . FACW
Anthoxanthum odoratum . . . FACU
Aruncus sylvester . . . . NOL
Aster
  chilensis . . . . FAC
  curtus . . . . NOL
  hallii . . . . FAC
  subspicatus . . . . FACW
Athyrium filix-femina . . . . FAC
Azolla
  filiculoides . . . . OBL
  mexicana . . . . OBL
Balsamorhiza
  deltoidea . . . . NOL
  sagittata . . . . NOL
Beckmannia syzigachne . . . . . OBL
Betula glandulosa . . . . NA
Bidens
  cernua . . . . FACW+
  frondosa . . . . FACW+
  tripartita . . . . FACW
Brasenia schreberi . . . . OBL
Briza
  maxima . . . . NOL
  minor . . . . FAC
Brodiaea elegans . . . . FACU

Bromus
  briziformis . . . . NOL
  carinatus . . . . NOL
  commutatus . . . . NOL
Calamagrostis canadensis . . FACW
Callitriche
  hermaphroditica . . . . OBL
  heterophylla . . . . OBL
  stagnalis . . . . OBL
  verna . . . . OBL
Calocedrus decurrens . . . . . . NOL
Caltha asarifolia . . . . NOL
Camassia
  leichtlinii . . . . FACW-
  quamash . . . . FACW
Cardamine
  occidentalis . . . . FACW+
  oligosperma . . . . FAC
  penduliflora . . . . OBL
Carex
  amplifolia . . . . FACW+
  aperta . . . . FACW
  aquatilis var. dives . . . . OBL
  cusickii . . . . OBL
  densa . . . . OBL
  deweyana . . . . FACU
  feta . . . . FACW
  hendersonii . . . . FAC
  lanuginosa . . . . OBL
  leporina . . . . FACW
  lyngbyei . . . . OBL
  mertensii . . . . FAC
  obnupta . . . . OBL
  pachystachya . . . . FAC
  retrorsa . . . . FAC
  scoparia . . . . FACW
  stipata . . . . FACW+
  tumulicola . . . . NOL
  unilateralis . . . . FACW
  utriculata . . . . OBL
  vesicaria var. major . . . . . . OBL
  vulpinoidea . . . . OBL
Castilleja tenuis . . . . FACU-
Centaurium
  erythraea . . . . FAC-
  muhlenbergii . . . . FACW
Ceratophyllum demersum . . . OBL
Chamaecyparis
  lawsoniana . . . . FACU+
  nootkatensis . . . . FAC
Chara sp. . . . . NOL
Cicendia quadrangularis . . . . . NOL
Cicuta douglasii . . . . OBL

Claytonia
  perfoliata . . . . FAC
  sibirica . . . . FAC
Conium maculatum . . . . FAC+
Cornus
  nuttallii . . . . NOL
  sericea . . . . FACW
Corylus cornuta . . . . FACU
Crassula aquatica . . . . OBL
Crataegus
  douglasii . . . . FAC
  var. suksdorfii . . . . NOL
  monogyna . . . . FACU+
  oxyacantha . . . . NOL
Crepis
  capillaris . . . . FACU
  setosa . . . . FACU
Cyperus aristatus . . . . OBL
Dactylis glomerata . . . . FACU
Danthonia californica . . . . . . FACU
Deschampsia
  cespitosa . . . . FACW
  danthonioides . . . . FACW-
  elongata . . . . FACW-
Dianthus armeria . . . . NOL
Dipsacus
  fullonum . . . . FAC
  sativus . . . . NOL
Downingia
  elegans . . . . OBL
  yina . . . . OBL
Echinochloa crus-galli . . . . . FACW
Egeria densa . . . . OBL
Elatine spp. . . . . OBL
Eleocharis
  acicularis . . . . OBL
  ovata . . . . OBL
  palustris . . . . OBL
Elodea canadensis . . . . OBL
Epilobium
  angustifolium . . . . FACU+
  brachycarpum . . . . FACW
  ciliatum
    ssp. glandulosum . . . . . FACW-
    ssp. watsonii . . . . FACW-
  densiflorum . . . . FACW-
  glaberrimum
    ssp. glaberrimum . . . . . FACW
  torreyi . . . . FACW
Equisetum
  arvense . . . . FAC
  fluviatile . . . . OBL
  telmateia ssp. braunii . . . . FACW

Erigeron decumbens . . . . NOL
Eriophyllum lanatum . . . . NOL
Eryngium petiolatum . . . . OBL
Festuca
    arundinacea . . . . FAC-
    pratensis . . . . FACU+
    rubra . . . . FAC+
Fontinalis antipyretica . . . . . . NOL
Fraxinus latifolia . . . . FACW
Galium
    aparine . . . . FACU
    cymosum . . . . FACW
    parisiense . . . . NOL
    trifidum var. pacificum . . FACW+
Geum macrophyllum . . . . . FACW-
Glyceria
    borealis . . . . OBL
    elata . . . . FACW+
    grandis . . . . OBL
    occidentalis . . . . OBL
    striata . . . . OBL
Gnaphalium
    palustre . . . . FAC+
    uliginosum . . . . FAC+
Grindelia integrifolia . . . . . . FACW
Helenium autumnale . . . . . .FACW
Heracleum
    lanatum . . . . FAC+
    mantegazzianum . . . . NOL
Hieracium spp. . . . . NOL
Holcus
    lanatus . . . . FAC
    mollis . . . . FACU
Holodiscus discolor . . . . NOL
Hordeum
    brachyantherum . . . . FACW-
    depressum . . . . FACW
    jubatum . . . . FAC
Howellia aquatilis . . . . OBL
Hydrocotyle ranunculoides . . . OBL
Hypericum
    anagalloides . . . . OBL
    formosum . . . . FAC-
Hypochaeris radicata . . . . . . FACU
Iris pseudacorus . . . . OBL
Isoetes
    bolanderi . . . . OBL
    echinospora . . . . OBL
    howellii . . . . OBL
    lacustris . . . . OBL
    nuttallii . . . . OBL
Juncus
    acuminatus . . . . OBL
    articulatus . . . . OBL
    balticus . . . . FACW+
    bolanderi . . . . OBL
    bufonius . . . . FACW+
    covillei . . . . FACW
    effusus . . . . FACW
    ensifolius . . . . FACW

hemiendytus . . . . FACW+
marginatus . . . . NOL
nevadensis . . . . FACW
oxymeris . . . . FACW+
patens . . . . FACW
tenuis . . . . FACW-
Lactuca
    serriola . . . . FAC
    tatarica ssp. pulchella . . . . . FAC
Lapsana communis . . . . NOL
Ledum glandulosum . . . . . FACW+
Leersia oryzoides . . . . OBL
Lemna
    minor . . . . OBL
    trisulca . . . . OBL
Leontodon autumnalis . . . . . . FAC
Limosella aquatica . . . . OBL
Lomatium
    bradshawii . . . . FACW
    triternatum . . . . NOL
    utriculatum . . . . NOL
Lonicera involucrata . . . . FAC
Lotus
    corniculatus . . . . FAC
    formosissimus . . . . FACW+
    pinnatus . . . . FACW
Ludwigia palustris . . . . OBL
Lupinus
    polyphyllus . . . . FAC+
    rivularis . . . . FAC
Lycopus
    americanus . . . . OBL
    uniflorus . . . . OBL
Lysichiton americanum . . . . . OBL
Lysimachia nummularia . . . . FACW
Lythrum
    hyssopifolium . . . . OBL
    portula . . . . OBL
    salicaria . . . . OBL
Madia glomerata . . . . FACU-
Malus fusca . . . . FACW
Mentha
    arvensis . . . . FACW-
    piperita . . . . FACU+
    pulegium . . . . OBL
    spicata . . . . OBL
Menyanthes trifoliata . . . . OBL
Mimulus
    dentatus . . . . OBL
    guttatus . . . . OBL
    moschatus . . . . FACW+
Montia
    fontana . . . . OBL
    howellii . . . . FACW-
    linearis . . . . NOL
Myosotis
    discolor . . . . FACW
    scorpioides . . . . FACW
    laxa . . . . OBL

Myriophyllum
    aquaticum . . . . OBL
    hippuroides . . . . OBL
    spicatum . . . . OBL
Najas
    flexilis . . . . OBL
    guadalupensis . . . . OBL
Navarretia intertexta . . . . FACW
Nitella sp. . . . . NOL
Nuphar lutea ssp. polysepala . OBL
Nymphaea odorata . . . . OBL
Oemleria cerasiformis . . . . . . FACU
Oenanthe sarmentosa . . . . . . OBL
Orthocarpus bracteosus . . . . . NOL
Panicum
    acuminatum . . . . FACW
    capillare . . . . FACU+
Parentucellia viscosa . . . . FAC-
Paspalum distichum . . . . . . FACW
Petasites frigidus
    var. palmatus . . . . FACW-
Phalaris
    aquatica . . . . FACU+
    arundinacea . . . . FACW+
Phleum pratense . . . . FAC-
Phragmites australis . . . . . FACW+
Physocarpus capitatus . . . . FACW-
Plagiobothrys
    figuratus . . . . FACW
    scouleri . . . . FACW
Plantago
    lanceolata . . . . FAC
    major . . . . FACU+
Poa
    annua . . . . FAC
    compressa . . . . FACU
    palustris . . . . FAC
    pratensis . . . . FAC
    trivialis . . . . FACW
Polygonum
    amphibium . . . . OBL
    coccineum . . . . OBL
    hydropiper . . . . OBL
    hydropiperoides . . . . OBL
    lapathifolium . . . . FACW
    persicaria . . . . FACW
    punctatum . . . . OBL
Populus balsamifera
    var. trichocarpa . . . . FAC
Potamogeton
    amplifolius . . . . OBL
    crispus . . . . OBL
    epihydrus . . . . OBL
    foliosus . . . . OBL
    gramineus . . . . OBL
    natans . . . . OBL
    nodosus . . . . OBL
    pectinatus . . . . OBL
    praelongus . . . . OBL

Potentilla
   anserina ssp. pacifica ..... OBL
   gracilis .... FAC
   palustris .... OBL
Prunella vulgaris .... FACU+
Pseudoroegneria spicata ... FACU-
Pseudotsuga menziesii ..... FACU
Psilocarphus elatior .... FACW
Puccinellia spp. .... NOL
Pyrus
   communis .... NOL
   fusca .... FAC+
   malus .... FACW
Quercus garryana .... NOL
Ranunculus
   alismifolius .... FACW
   aquatilis .... OBL
   flabellaris .... OBL
   flammula .... FACW
   lobbii .... OBL
   occidentalis .... FAC
   orthorhynchus .... FACW-
   repens .... FACW
   sceleratus .... OBL
   uncinatus .... FAC-
Ricciocarpos natans .... NOL
Rorippa
   curvisiliqua .... FACW+
   islandica .... OBL
   nasturtium-aquaticum ..... OBL
Rosa
   eglanteria .... FACW
   multiflora .... NI
   nutkana .... FAC
   pisocarpa .... FAC
Rubus
   parviflorus .... FACU+
   spectabilis .... FAC+
Rumex
   acetosella .... FACU+
   conglomeratus .... FACW-
   crispus .... FACW
   obtusifolius .... FAC
   salicifolius .... FACW
Sagittaria latifolia .... OBL

Salix
   fluviatilis .... OBL
   geyeriana .... FACW+
   hookeriana .... FACW-
   lucida ssp. lasiandra .... FACW+
   rigida .... OBL
   scouleriana .... FAC
   sessilifolia .... FACW
   sitchensis .... FACW
Saxifraga
   ferruginea .... FAC
   integrifolia .... FACW
   nuttallii .... OBL
   oregana .... FACW+
Scirpus
   acutus .... OBL
   cyperinus .... OBL
   microcarpus .... OBL
   tabernaemontani .... OBL
Senecio integerrimus ...... FACU
Sidalcea
   campestris .... NI
   nelsoniana .... FAC
Sisyrinchium
   hitchcockii .... NOL
   idahoense .... FACW
Sium suave .... OBL
Sonchus
   arvensis .... FAC-
   oleraceus .... UPL
Sparganium
   emersum .... OBL
   eurycarpum .... OBL
Spiraea douglasii .... FACW
Spirodela polyrhiza .... OBL
Stachys
   cooleyae .... FACW
   mexicana .... FACW
   palustris .... FACW+
   rigida .... FACW-
Stellaria media .... FACU
Symphoricarpos
   albus .... FACU
   mollis .... NOL
Taraxacum officinale .... NOL

Taxus brevifolia .... FACU-
Thuja plicata .... FAC
Tragopogon dubius .... NOL
Triteleia hyacinthina .... NOL
Typha
   angustifolia .... OBL
   latifolia .... OBL
Urtica dioica .... FAC+
Utricularia
   gibba .... OBL
   intermedia .... OBL
   macrorhiza .... OBL
   minor .... OBL
Vaccinium caespitosum .... FACU
Vallisneria americana .... OBL
Veratrum
   californicum .... FACW+
   viride .... FACW
Veronica
   americana .... OBL
   anagallis-aquatica .... OBL
   arvensis .... FACU
   cusickii .... FAC
   peregrina .... OBL
   scutellata .... OBL
   wormskjoldii .... FAC
Viburnum
   edule .... NA
   ellipticum .... NOL
Wolffia
   borealis .... OBL
   columbiana .... OBL
Wyethia angustifolia .... FACU
Zannichellia palustris .... OBL
Zigadenus venenosus ...... FAC

APPENDIX 3

Endemic, native and introduced species.

Endemic species

Aster curtus
Aster hallii
Cardamine penduliflora
Erigeron decumbens
Eryngium petiolatum
Lomatium bradshawii
Sidalcea campestris
Sidalcea nelsoniana

Native species

Alisma gramineum
Alisma plantago-aquatica
Alnus rubra
Alopecurus aequalis
Alopecurus geniculatus
Amelanchier alnifolia
Angelica genuflexa
Aster chilensis
Aster subspicatus
Azolla mexicana
Beckmannia syzigachne
Bidens cernua
Brasenia schreberi
Callitriche hermaphroditica
Callitriche heterophylla
Callitriche verna
Camassia quamash
Carex aperta
Carex densa
Carex deweyana
Carex feta
Carex lanuginosa
Carex obnupta
Carex stipata
Carex unilateralis
Carex vesicaria
var. major
Centaurium muhlenbergii
Ceratophyllum demersum
Cicuta douglasii
Claytonia sibirica
Cornus sericea
Crataegus douglasii
Danthonia californica
Deschampsia cespitosa
Downingia elegans
Eleocharis acicularis

Eleocharis ovata
Eleocharis palustris
Elodea canadensis
Epilobium ciliatum
ssp. watsonii
Epilobium densiflorum
Epilobium torreyi
Equisetum arvense
Eriophyllum lanatum
Festuca rubra
Fontinalis antipyretica
Fraxinus latifolia
Galium trifidum
var. pacificum
Geum macrophyllum
Glyceria elata
Glyceria occidentalis
Gnaphalium palustre
Grindelia integrifolia
Heracleum lanatum
Hordeum
brachyantherum
Howellia aquatilis
Hydrocotyle
ranunculoides
Hypericum anagalloides
Isoetes nuttallii
Juncus acuminatus
Juncus articulatus
Juncus bufonius
Juncus covillei
Juncus effusus
Juncus ensifolius
Juncus nevadensis
Juncus oxymeris
Juncus patens
Juncus tenuis

Leersia oryzoides
Lemna minor
Limosella aquatica
Lotus formosissimus
Lotus pinnatus
Lupinus polyphyllus
Lycopus americanus
Lysichiton americanum
Lythrum hyssopifolium
Mentha arvensis
Mimulus guttatus
Montia fontana
Myosotis laxa
Myriophyllum
hippuroides
Najas guadalupensis
Nuphar lutea
ssp. polysepala
Oenanthe sarmentosa
Orthocarpus bracteosus
Panicum capillare
Physocarpus capitatus
Plagiobothrys figuratus
Polygonum amphibium
Polygonum
hydropiperoides
Populus balsamifera
var. trichocarpa
Potamogeton gramineus
Potamogeton epihydrus
Potamogeton foliosus
Potamogeton natans
Potamogeton praelongus
Potentilla gracilis
Pyrus fusca
Ranunculus alismifolius
Ranunculus aquatilis

Ranunculus flammula
Ranunculus occidentalis
Ranunculus
orthorhynchus
Ricciocarpos natans
Rorippa curvisiliqua
Rorippa islandica
Rorippa nasturtium-
aquaticum
Rosa nutkana
Rubus spectabilis
Sagittaria latifolia
Salix fluviatilis
Salix geyeriana
Salix hookeriana
Salix lucida ssp. lasiandra
Salix scouleriana
Salix sessilifolia
Salix sitchensis
Saxifraga oregana
Scirpus microcarpus
Scirpus tabernaemontani
Sisyrinchium idahoense
Sparganium emersum
Spiraea douglasii
Spirodela polyrhiza
Stachys cooleyae
Symphoricarpos albus
Thuja plicata
Triteleia hyacinthina
Typha latifolia
Utricularia macrorhiza
Veratrum californicum
Veronica americana
Veronica scutellata
Wolffia borealis
Wyethia angustifolia

Introduced Species

Agrostis capillaris
Alisma lanceolatum
Alopecurus pratensis
Anthoxanthum odoratum
Bidens frondosa
Briza minor
Bromus briziformis
Callitriche stagnalis
Centaurium erythraea
Crataegus monogyna
Crataegus oxyacantha
Dianthus armeria
Dipsacus fullonum

Dipsacus sativus
Echinochloa crus-galli
Egeria densa
Festuca arundinacea
Festuca pratensis
Galium parisiense
Heracleum
mantegazzianum
Holcus lanatus
Iris pseudacorus
Juncus marginatus
Lotus corniculatus

Ludwigia palustris
Lysimachia nummularia
Lythrum portula
Lythrum salicaria
Mentha piperita
Mentha pulegium
Mentha spicata
Myosotis discolor
Myosotis scorpioides
Myriophyllum aquaticum
Myriophyllum spicatum
Nymphaea odorata

Parentucellia viscosa
Phalaris arundinacea
Poa palustris
Poa pratensis
Polygonum hydropiper
Potamogeton crispus
Pyrus communis
Pyrus malus
Rosa eglanteria
Rumex crispus
Vallisneria americana
Veronica peregrina

APPENDIX 4

Marginal, transitional, disturbed-site and upland species.

Marginal (M): a wetland species that can (but does not always) indicate marginal wetland or upland conditions.

Transitional (T): a species that often (but not always) indicates a transition from wetland to upland, or an upland/wetland boundary.

Disturbed-site (D): a species that often (but not always) indicates or tends to favor disturbed conditions.

Upland (U): a species that most often (but not always) indicates or tends to prefer non-wetland conditions.

Species	M	T	D	U
Acer circinatum	M	T		U
Agrostis				
alba				
var. *alba*	M	T		
var. *palustris*	M	T		
diegoensis	M	T		
idahoensis	M	T		
stolonifera	M	T		
Amelanchier alnifolia	M	T		U
Anthoxanthum odoratum	M	T	D	
Aruncus sylvester	M	T		U
Aster chilensis	M	T		
Balsamorhiza				
deltoidea				U
sagittata				U
Briza				
maxima				U
minor	M	T	D	
Bromus				
briziformis				U
commutatus	M		D	U
Carex				
mertensii	M			
retrorsa	M	T		
tumulicola	M		D	
Castilleja tenuis	M	T		
Centaurium muhlenbergii	M	T		
Ceratophyllum demersum			D+	
Chamaecyparis				
lawsoniana	M	T		U
nootkatensis	M			U
Claytonia perfoliata	M	T	D	U
Corylus cornuta	M	T		U
Crataegus				
douglasii var. *suksdorfii*	M	T	D	U
monogyna	M	T	D	U
oxyacantha	M	T	D	U
Crepis				
capillaris			D	U
setosa			D	U
Dactylis glomerata	M		D	U
Danthonia californica	M	T		
Dipsacus fullonum	M		D	
Echinochloa crus-galli	M		D	
Egeria densa			D+	
Epilobium				
brachycarpum	M		D	U
ciliatum ssp. *watsonii*	M		D	U
densiflorum	M	T	D	
Equisetum				
arvense	M		D	U
telmateia ssp. *braunii*	M		D	
Festuca				
arundinacea	M	T	D	U
pratensis	M			U
rubra	M			
Galium aparine	M		D	U
Gnaphalium				
palustre	M		D	
uliginosum	M		D	
Grindelia integrifolia			D	
Hieracium spp.	M			U
Holcus				
lanatus	M	T	D	
mollis				U
Holodiscus discolor	M			U
Hordeum jubatum	M	T	D	
Juncus				
bufonius	M		D	
marginatus	M	T	D	
tenuis	M	T	D	
Lactuca				
serriola			D	U
tatarica ssp. *pulchella*			D	U

Lapsana communis .D U
Leontodon autumnalis . . . . .M   U
Lotus corniculatus . . . . .M   T
Ludwigia palustris .D
Lupinus
    polyphyllus . . . . .M   T   U
    rivularis . . . . .M   T   U
Lythrum
    hyssopifolium . . . . .D+
    portula . . . . .D
    salicaria . . . . .D+
Madia glomerata . . . . .M   T   D   U
Mentha pulegium . . . . .D+
Myosotis discolor . . . . .M   U
Myriophyllum aquaticum . . . . .D+
Navarretia intertexta . . . . .M   T   D
Oemleria cerasiformis . . . . .M   U
Orthocarpus bracteosus . . . . .M   T   D   U
Panicum
    acuminatum . . . . .M   T
    capillare . . . . .M   T   D
Parentucellia viscosa . . . . .M   T   D   U
Petasites frigidus var. palmatus . . . . . .M   T   U
Phalaris
    aquatica . . . . .M   D+
    arundinacea . . . . .D+
Phleum pratense . . . . .M   D   U
Physocarpus capitatus . . . . .M   U
Plantago
    lanceolata . . . . .M   T   D   U
    major . . . . .M   T   D   U
Poa
    annua . . . . .M   T   D   U
    compressa .M
    pratensis . . . . .M   T   D   U

Populus balsamifera var. trichocarpa . .M
Prunella vulgaris . . . . .M   T   D   U
Pseudotsuga menziesii . . . . .U
Pyrus
    communis .M
    fusca . . . . .M
Quercus garryana . . . . .U
Ranunculus repens . . . . .M   D
Rosa multiflora . . . . .D+
Senecio integerrimus .D   U
Sidalcea
    campestris . . . . .M
    nelsoniana .M
Stellaria media . . . . .M   D   U
Symphoricarpos
    albus . . . . .M   T   U
    mollis . . . . .U
Taraxacum officinale . . . . .D+ U
Taxus brevifolia . . . . .M   U
Tragopogon dubius . . . . .M   D   U
Triteleia hyacinthina . . . . .M
Urtica dioica . . . . .M   T
Vaccinium caespitosum . . . . .M   U
Veronica arvensis . . . . .M   D   U
Viburnum
    edule . . . . .M
    ellipticum . . . . .M
Wyethia angustifolia . . . . .M
Zigadenus venenosus . . . . .M

ILLUSTRATED GLOSSARY

achene: a small, dry, 1-seeded, non-splitting fruit

acuminate: tapered to a sharp point, with the edges somewhat concave

aggregate fruit: a fruit-like structure composed of several separate, fleshy, single-seeded fruits (e.g., raspberry, blackberry)

alkaline: rich in soluble mineral salts (chlorides, sulfates, carbonates and bicarbonates of sodium, potassium, magnesium and calcium) that can neutralize acids

acuminate leaf

alluvial: composed of deposits laid down by flowing water, such as on a floodplain or at a rivermouth

alternate: situated singly at each node (cf. *opposite*)

amphibious: adapted for living or growing both on land and in water (cf. *aquatic, semi-aquatic, terrestrial*)

annual: a plant that completes its life cycle and dies in a single year (cf. *biennial, perennial*)

anther: the pollen-producing structure at the top of a stamen

anthesis: the period during which a flower is fully open and functional

berry

aquatic: living in waterlogged soil or wholly or partly submerged in water (cf. *amphibious, semi-aquatic, terrestrial*)

ascending: growing obliquely upwards (stems); directed obliquely forwards in respect to the organ to which it is attached (parts of a plant)

auricle: a small lobe or appendage, often projecting from the base of an organ

awn: a slender bristle, usually at the tip of a structure, as in achenes, grass flowers, bracts and scales

axil: the angle between a leaf and the stem to which it is attached

axis: a central longitudinal line about which the parts of a plant are arranged

basal: situated at or near the base of a structure

berry: a soft, fleshy fruit, usually spherical and many-seeded, and with a tough outer skin (e.g., blueberry, snowberry, grape)

biennial: a plant that completes its life cycle and dies in two years; the flowers and fruits are usually produced only in the second year (cf. *annual, perennial*)

bilabial: two-lipped

blade: the broad, flat part of a leaf or petal

brackish: slightly salty

bract: a reduced or specialized leaf associated with, but not part of, a flower or flower cluster; a small leaf-like structure or a scale that bears the reproductive organs in lower vascular plants

bryophyte: any plant of the group Bryophyta, comprising the liverworts and mosses

budding: a form of asexual reproduction in which a new individual develops from an outgrowth of a mature plant

bulb: a short, vertical underground shoot with modified leaves or thickened leaf bases developed as food-storage organs (cf. *corm*)

calyx: all the sepals of a flower, collectively, which form the outer whorl of the flower (cf. *corolla*)

campanulate flower

campanulate: bell-shaped

capsule: a dry fruit that splits open at maturity and is composed of more than one carpel (cf. *follicle*); in mosses and liverworts, the spore-containing sac

catkin: a dense, spike-like arrangement of small flowers, usually of one sex, that lack petals but usually have surrounding bracts, in the willow, oak or birch family (also called an *ament*)

caudex: a short, more or less vertical, often woody stem at or just below the surface of the ground, from which new stems arise each year; it may produce roots like a *rhizome*

clone: a group of genetically identical individuals arising by vegetative multiplication from a single individual

Leaf Characteristics

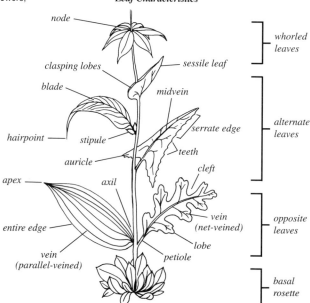

node

clasping lobes

blade

hairpoint

auricle

apex

entire edge

vein
(parallel-veined)

stipule

axil

sessile leaf

midvein

serrate edge

teeth

cleft

vein
(net-veined)

lobe

petiole

whorled leaves

alternate leaves

opposite leaves

basal rosette

composite: a plant of the family Asteraceae (Compositae), characterized by a head-like flower cluster that looks like a single flower

compound: divided into similar smaller parts, such as a leaf into leaflets (cf. *simple*)

corm: a flattened, often disk-like underground stem that is thickened as a food-storage organ, without prominently thickened leaves (cf. *bulb*)

corolla: the petals of a flower, collectively, which form the inner whorl of the flower (cf. *calyx*)

corymb: a flat- or round-topped flower cluster in which the outer flowers bloom first

cosmopolitan: found all over the world

creeping: growing along (or beneath) the surface of the ground and producing roots at intervals, usually at the nodes

corymb

crisped: with irregularly-curled, wavy edges

culm: the aerial stem of a grass or sedge

cultivar: a cultivated variety of a plant species

cutting: a detached portion of a living plant, such as a leaf or bud, that can produce a new daughter plant if grown in soil or in a suitable culture medium

cyme: a broad class of flower clusters in which the flower at the tip of the cluster, and often the flowers at the tips of any branches, bloom first (cf. *raceme*)

deciduous: falling off after the completion of normal function (cf. *persistent*); said of trees and shrubs that usually lose their leaves at the approach of winter

decumbent: reclining or lying flat on the ground, but with ascending tips

deflation plain: a broad interdune area that is wind-scoured to the level of the summer water table

dentate: toothed, as in a leaf edge, with rounded, tooth-like projections (cf. *serrate*)

cyme

desiccation: becoming dried up

dichotomous: forking more or less regularly into two branches of about equal size

dicotyledon: a seed plant with two cotyledons (seed leaves) in the embryo; most have leaves with net veins; includes most families of flowering plants in our region (cf. *monocotyledon*)

dimorphic: of two forms

disk flower: one of the many small tubular flowers found in the central disk of a head-like flower cluster in the aster family (cf. *ray flower*)

dissected: deeply divided (often repeatedly) into small or slender parts or lobes

dissected leaf

diurnal: of the daytime (diurnal flowers are open during the day)

divided: cut into separate parts, as a leaf that is cut to its midrib or base

drupe: a fleshy fruit with a central stony pit that permanently encloses the seed (e.g., cherry, plum, peach, olive)

ecotone: a transition zone separating two ecological communities

ecotype: the individuals of a species that are adapted to a particular habitat; sometimes classified into a variety or subspecies

elliptical: shaped like an ellipse; oval or oblong with rounded ends and widest in the middle

emergent: coming out from; often refers to plants that have their bases submerged in water (cf. *submerged*)

endemic: growing only in a particular region; the Willamette Valley wet prairie community has several endemic species, some of which are nearly extinct and receive protection under the Federal Threatened and Endangered Species Act (cf. *introduced, native*)

entire: not toothed or otherwise cut

eutrophic: rich in organic or mineral nutrients; refers especially to lakes and ponds where the resultant growth and decay of algae and other plants significantly depletes oxygen levels (cf. *oligotrophic*)

extirpated: locally extinct; without any living representatives in a region that was formerly part of its range, but still present in other places

fibrous: formed from or full of fibers

filament: part of a stamen; the stalk that supports the anther

filiform: very slender, thread-like

flaccid: weak and lax; hardly if at all capable of supporting its own weight

fleshy: thick and juicy

floodplain: an area adjacent to a stream or river that is subject to repeated flooding

floret: a tiny flower, usually part of a cluster; used especially to describe the specialized flower of a grass

Parts of a Composite Flowerhead

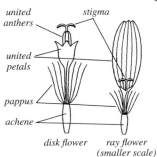

united anthers
stigma
united petals
pappus
achene
disk flower
ray flower
(smaller scale)

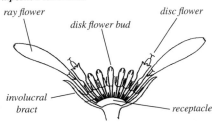

ray flower
disk flower bud
disc flower
involucral bract
receptacle

section through a flowerhead

follicle: a dry, single-carpel fruit that splits at maturity along one side only (cf. *capsule*)

forb: a broad-leaved, non-woody plant that dies back to the ground each year

fragmentation: a form of asexual reproduction in which new individuals arise from parts broken off the parent plant

fruit: a ripened ovary, together with any other structures that ripen with it as a unit

gall: an abnormal localized swelling or outgrowth produced by a plant as a result of attack by a parasite

glabrous: hairless

gland: a bump, appendage or depression on a plant's surface that produces a sticky or greasy, viscous fluid

glandular: having glands

glaucous: covered with a waxy grayish-blue powder that can be rubbed off

globose: shaped like a sphere

glume: one of two empty bracts at the base of a grass spikelet

head: a cluster of stalkless flowers crowded closely together at the tip of a peduncle

herb: a plant, whether annual, biennial or perennial, with non-woody stems that die back to the ground at the end of the growing season

herbaceous: herb-like

hip: the fruit-like structure of roses, consisting of a large, red, cup-shaped, hollow receptacle that contains flask-like achenes

incised: sharply (and often irregularly) cut along an edge, sometimes deeply so

inflated: having a bulging form and hollow interior

inflorescence: a flower cluster; more correctly, the arrangement of flowers on the axis

insectivorous: feeding on insects

interrupted: discontinuous, broken by regions of bare axis

introduced: non-native to the area in which it occurs; brought to a region by human activity (cf. *endemic, native*)

inundation: a condition in which water temporarily or permanently covers a land surface (cf. *saturation*)

invertebrate: an animal without a backbone or spinal column; insects and crustaceans are the most common invertebrates in freshwater habitats

involucre: a set of bracts below a flower cluster, especially in the aster family

irregular flower: a flower with petals (or less often sepals) that are dissimilar in form or orientation

keel-shaped: with a sharp or conspicuous longitudinal ridge

keel: the two partly fused lower petals of many pea family flowers

lance-shaped: narrowly elliptical, longer than broad, broadest near base and tapering at the tip

lateral: on the side

lax: loose, limp

leaf scar: a scar left on a twig when a leaf falls.

leaflet: one part or division of a compound leaf

legume: the fruit of a pea family plant, composed of a single carpel, typically dry and splitting down both seams; a plant of the pea family

lemma: the lower of the two bracts immediately enclosing an individual grass flower

linear: line-shaped; very long and narrow with essentially parallel sides

Leaf Shapes

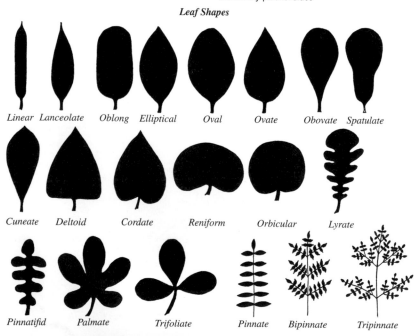

Linear Lanceolate Oblong Elliptical Oval Ovate Obovate Spatulate

Cuneate Deltoid Cordate Reniform Orbicular Lyrate

Pinnatifid Palmate Trifoliate Pinnate Bipinnate Tripinnate

lip: a projection or expansion of something, such as the lower petal of an orchid or violet flower

lobe: a projecting segment of an organ, larger than a tooth or auricle

macrophyte: any plant, especially aquatic plant, large enough to be discerned by the naked eye

margin: an edge

mesic: moist, neither very wet nor very dry

midrib: the main, central rib of a leaf

midvein: the main, central vein of a leaf

mitigation: a plan and execution (whether successful or not) of compensatory replacement, either by creating new wetlands or enhancing existing wetlands, of certain officially recognized and quantifiable wetland functions and values that will be lost to a wetland that is permitted to fill

monocotyledon: a seed plant with one cotyledon (seed leaf) in the embryo; most have leaves with parallel veins; includes many aquatic plants in our region (cf. *dicotyledon*)

monoculture: a single species comprising the total vegetation of a tract of land

monotypic: containing only a single type, as a genus with only one species

mud flat: a stretch of muddy land left uncovered at low tide; a muddy shore along a non-tidal river

native: naturally occurring in a region, not introduced by human activity (cf. *endemic, introduced*)

naturalized: thoroughly established and adapted to the environment of a region, but originally coming from another area

nerve: a prominent longitudinal line or vein in a leaf, petal or other organ

node: a joint or section of a stem or axis where a branch, root or leaf is (or has been) attached

noxious weed: an introduced plant considered harmful to animals or the environment

nut: a hard, dry, usually one-seeded fruit that does not open at maturity; larger and thicker-walled than an achene

nutlet: a small nut; a very thick-walled achene

oblanceolate: lance-shaped, but broadest nearer the tip and more sharply narrowed near the base

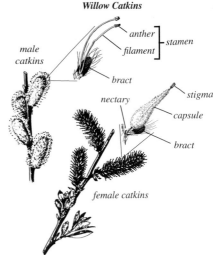

Willow Catkins

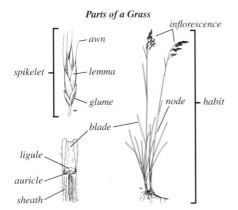

Parts of a Grass

oblong: more or less rectangular with rounded corners

obtuse: blunt, with the sides coming together at an angle of more than 90°

offset: situated nearly but not quite opposite along an axis; a short, prostrate, primarily propagative shoot originating near the ground

oligotrophic: relatively poor in plant nutrients and containing abundant oxygen in the deeper parts (cf. *eutrophic*)

opposite: situated directly across from each other at the same node (cf. *alternate*)

oval: broadly elliptical

ovary: the structure at the base of a pistil that contains the young, undeveloped seeds

ovate: shaped like a long section through a hen's egg, broadest below the middle, abruptly narrowed at the tip

palmate: divided into three or more lobes or leaflets that diverge from a common point, like fingers of a hand (cf. *pinnate*)

panicle: a branched flower cluster, common in grass species

pedicel: the stalk of an individual flower in an inflorescence

peduncle: the stalk of a inflorescence or of a solitary flower

peltate: shield-like, as a leaf with the stalk attached in the center of the blade

perennial: a plant that lives for more than two years (cf. *annual, biennial*)

perianth: the sepals and petals of a flower, collectively

perigynium [perigynia]**:** a special bract that encloses an achene (in a *Carex* sedge)

persistent: remaining attached after the normal function has been completed (cf. *deciduous*)

petal: a member of the inner ring of modified flower leaves, usually white or brightly colored

petiole: a leaf stalk, bearing one or more blades

pinnate: feather-like; with a row of leaflets on either side of a central stalk, like the barbs of a feather (cf. *palmate*)

pistil: the female organ of a flower, usually consisting of an ovary, style and stigma

pistillate: with a pistil; female; usually applied to flowers with pistils only (lacking stamens)

pith: the soft, spongy tissue in the center of the stems and branches of certain plants

pod: a dry fruit that splits open at maturity

pome: a fruit with a core (e.g., apple, pear)

prickle: a spine or thorn-like growth

propagation: reproduction by natural processes

prostrate: lying flat on the ground

raceme: an unbranched cluster of stalked flowers arranged along a common axis, blooming from the bottom upwards

rachis: a main axis, as in a compound leaf

radiate: spread outwards around a common center, like the spokes of a wheel

ray: a branch of an umbel

ray flower: one of several flowers, each with a flat-tened, strap-like corolla, that radiate around the central disk of aster family flower heads (cf. *disk flower*)

reflexed bracts subtending a flower

reflexed: abruptly bent downwards or backwards

rhizoid: a fine hair-like extension of a cell forming an anchoring or absorbing organ, as in liverworts

rhizomatous: having rhizomes

rhizome: an underground, often lengthened stem; distinguished from a root by the presence of nodes and buds or scale-like leaves

rib: one of the main longitudinal veins of a leaf or other organ

rootstock: a vertical rhizome

rosette: a cluster of organs (usually leaves) arranged in a circle or disk, often at the base of a structure

samara: a dry, usually one-seeded, winged fruit (e.g., maple and ash fruits)

saturation: a condition in which all easily drained voids (pores) between soil particles are temporarily or permanently filled with water; significant saturation is considered to last for one week or more during the growing season (cf. *inundation*)

savannah: a tract of low-lying, damp or marshy ground

scale: any small, thin or flat structure

scape: a leafless flower stalk arising from near the ground

scorpioid: curled or coiled inwards, uncoiling later

seed: a ripened ovule

segment: one part of a flower, leaf, etc. which is divided, but not compound.

scorpioid inflorescence

semi-aquatic: adapted for living or growing in or near water, but not entirely aquatic (cf. *amphibious, semi-aquatic, terrestrial*)

sepal: a member of the outermost ring of modified flower leaves; usually green and more or less leafy in texture

serrate: sharply saw-toothed, with the teeth pointing forwards

sheath: a tubular part surrounding another part, often papery

sheath: an organ that partly or completely surrounds another organ, as the sheath of a grass leaf surrounds a stem

silicle: a short pod-like fruit of the mustard family, not much longer than wide and of various shapes, but often rounded

silique: a pod-like fruit of certain members of the mustard family, much longer than wide, which consists of two chambers that separate at maturity

siltation: becoming filled or blocked by the gradual accumulation of silt

simple: not divided or subdivided (cf. *compound*)

slough: a wet depression or pond

sorus [sori]: a cluster of small spore cases

spadix: a flowering spike with small, crowded flowers on a thickened, fleshy stalk; used only among mono-cotyledons

spathe: a large, usually solitary bract below and often enclosing a flower cluster (usually a spadix); used only among monocotyledons

spatulate: shaped like a spatula, or like a long section through a pear; rounded at the tip, rounded, but nar-rower at the base

spike: a flower cluster with stalkless or nearly stalk-less flowers arranged along an unbranched stalk, blooming from the bottom upwards

spikelet: a small or secondary spike in a panicle; a cluster of grass flowers and their bracts

sporangium: a spore case

spore: a one- to several-celled reproductive body produced in a capsule (mosses and liverworts) or sporangium (ferns and horsetails)

Parts of a Sedge Inflorescence

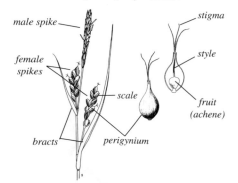

male spike

female spikes

scale

bracts

perigynium

stigma

style

fruit (achene)

sporophyte: the spore-bearing part or phase of a plant; in bryophytes composed of a foot, stalk and capsule; in vascular plants composed of the leafy plant

spur: a hollow appendage on a petal or sepal, usually functioning as a nectary

stamen: the pollen-bearing male organ of a flower, usually consisting of an anther and a filament

staminate: with stamens; male; usually applied to flowers with stamens only (lacking pistils)

stigma: the tip of a pistil, where the pollen lands

stipule: one of a pair of wing- or scale-like structures at the base of a leaf

stolon: a long, creeping stem on the surface of the ground

striations: fine, parallel ridges

style: the middle stalk of a pistil, connecting the stigma to the ovary; often remains as projection from tip of fruit

sub-: nearly, almost; as in subsessile, suborbicular, etc.

submerged: growing entirely underwater (cf. *emergent*)

substrate: a non-soil surface on which something grows

subtend: to be directly below and close to, as of a bract to a flower

succession: the process of change in plant community composition and structure over time

succulent: fleshy and juicy; describes plants that store reserves of water in fleshy stems or leaves

swale: a moist hollow or low place, often dominated by ash trees in our region

tepal: a sepal or petal, when these structures cannot be differentiated

terminal: at the end or top of a structure

terrestrial: living on or growing in the earth, land-based (cf. *amphibious, aquatic, semi-aquatic*)

thallus: a main plant body, not differentiated into stems and leaves, as in duckweeds and some liverworts

throat: the opening into the tube of a flower

trailing: lying flat on the ground, but not rooting

tuber: a thickening, usually at the end of a rhizome, serving in food storage and often also in reproduction; also sometimes loosely applied to tuberous roots

tuberous: thickened like a tuber

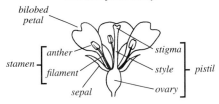

A Section through a Regular Flower with an Inferior Ovary

bilobed petal — anther — filament — sepal — stamen — stigma — style — ovary — pistil

tubular: a hollow structure with parallel edges; a long, slender corolla

turions: a small, bulb-like offset, as in some species of *Epilobium*

umbel: an often flat-topped flower cluster in which the flower stalks arise from a common point, much like the stays of a reversed umbrella

umbelet: one units of a compound umbel

vein: a strand of conducting tubes (vascular bundle), especially if visible externally, as in a leaf

vernal: coming, appearing or occurring in spring

wetland boundary: the point on the ground at which a shift from wetlands to non-wetlands occurs

whorl: a ring of three or more similar structures radiating from a common point on an axis (e.g., leaves around a node on a stem)

whorled: arranged in a whorl

zonation: the distribution of plants and animals into specific zones according to altitude, depth, etc., each characterized by its dominant species

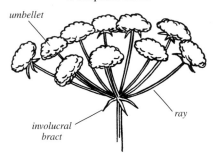

A Compound Umbel

umbellet — involucral bract — ray

spadix

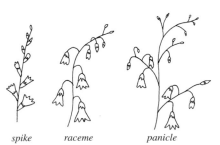

spike   *raceme*   *panicle*

REFERENCES

Abrams, L. 1940–59. *Illustrated Flora of the Pacific States.* 4 vols. Stanford University Press, Stanford, California.

Adamus, P.R., and L.T. Stockwell. 1982. *A Method for Wetland Functional Assessment.* Vol. 1, *Critical Review and Evaluation Concepts.* U.S. Department of Transportation. Federal Highway Administration.

Boss, T.R. 1983. Vegetation ecology and net primary productivity of selected freshwater wetlands in Oregon. Ph.D. dissertation. Oregon State University, Corvallis.

Bursche, E.M. 1968. *A Handbook of Water Plants.* Fredrick Warne & Co., London.

Christy, J.A. 1993. *Classification and Catalogue of Native Wetland Plant Communities in Oregon.* Oregon Natural Heritage Program, Portland.

Clark, L.J. 1974a. *Lewis Clark's Field Guide to Wild Flowers of the Sea Coast in the Pacific Northwest.* Gray's Publishing, Sidney, British Columbia.

———. 1974b. *Lewis Clark's Field Guide to Wild Flowers of Marsh and Waterway in the Pacific Northwest.* Gray's Publishing, Sidney, British Columbia.

———. 1976. *Wildflowers of the Pacific Northwest from Alaska to Northern California.* Ed. J.G. Trelawny. Gray's Publishing, Sidney, British Columbia.

Connelly, K.P. 1989. Ecological effects of fire in bottomland Willamette Valley prairies, with special emphasis on *Lomatium bradshawii* and *Erigeron decumbens,* two rare endemic plants. Study plan and progress report. The Nature Conservancy, Oregon Field Office, Portland.

Cook, C.D.K., B.J. Gut, E.M. Rix, J. Schneller and M. Seitz. 1974. *Water Plants of the World: A Manual for the Identification of the Genera of Freshwater Macrophytes.* Dr. W. Junk b.v., Publishers, The Hague.

Correll, D.S., and H.B. Correll. [1972] 1975. *Aquatic and Wetland Plants of SW United States.* 2 vols. U.S. Environmental Protection Agency. Reprint. Stanford University Press, Stanford, California.

Cowardin, L.M., V. Carter, F.C. Golet and E.T. LaRue. 1979. *Classification of Wetlands and Deepwater Habitats of the United States.* FWS/OBS-79/31. U.S. Fish & Wildlife Service.

Crawford, R.M.M. 1987. *Plant Life in Aquatic and Amphibian Habitats.* Blackwell Scientific Publications, Palo Alto, California.

Elias, T.S., and P.A. Dykeman. 1982. *A Field Guide to North American Edible Wild Plants.* Outdoor Life Books, New York.

Fairbrothers, D.E., and E.T. Moul. *Aquatic Vegetation of New Jersey.* Parts 1 & 2. Extension Bulletin No. 382. Extension Service College of Agriculture, Rutgers University, New Brunswick, New Jersey.

Fassett, N.C. 1957. *A Manual of Aquatic Plants.* University of Wisconsin Press, Madison.

Franklin, J.F., and C.T. Dyrness. 1973. *Natural Vegetation of Oregon and Washington.* U.S.D.A. Forest Service General Technical Report PNW-8.

Frenkel, R.E., and E.F. Heinitz. 1987. Composition and structure of Oregon ash (*Fraxinus latifolia*) forest in William L. Finley National Wildlife Refuge, Oregon. *Northwest Science* 61: 203–12.

Ganskopp, D.C. 1979. Plant communities and habitat types of the Meadow Creek Experimental Watershed. M.S. thesis. Oregon State University, Corvallis.

Gilkey, H.M. 1951. *Handbook of Northwest Flowering Plants.* 2nd ed. Binford and Mort, Portland, Oregon.

Glad, J.B., R. Mishaga and R.R. Halse. 1987. Habitat characteristics of *Sidalcea nelsoniana* Piper (Malvaceae) at Walker Flat, Yamhill County, Oregon. *Northwest Science* 61: 257–263.

Hall, M.J. 1976. Preliminary inventory and analysis: Santiam Bar Proposed Natural Area Preserve. Oregon State Land Board, Division of State Lands, Salem.

Hansen, H.P. 1941. Paleoecology of two peat deposits on the Oregon coast. *Oregon State Monographs, Studies in Botany* 3: 1–31.

Harris, S.W. 1954. An ecological study of the waterfowl of the Potholes area, Grant County, Washington. *American Midland Naturalist* 52: 403–432.

Heinitz, E.F. 1982. Vegetation ecology of *Fraxinus latifolia* communities in William L. Finley National Wildlife Refuge, Oregon. M.S. thesis. Oregon State University, Corvallis.

Hemstrom, M.A., S.E. Logan and W. Pavlat. 1985. *Preliminary Plant Association and Management Guide, Willamette National Forest.* U.S.D.A. Forest Service, Pacific Northwest Region.

———. 1987. Plant association and management guide, Willamette National Forest. U.S.D.A. Forest Service R6-Ecolo 257-B-86.

Henderson, J.A. 1970. Biomass and composition of the understory vegetation in some *Alnus rubra* stands in western Oregon. M.S. thesis. Oregon State University, Corvallis.

Hermann, F.J. 1975. *Manual of the Rushes* (Juncus spp.) *of the Rocky Mountains and Colorado Basin.* U.S.D.A. Forest Service General Technical Report Rm-18.

Hickman, J.C., ed. 1993. *The Jepson Manual: Higher Plants of California.* University of California Press, Berkeley.

Hinschberger, M.S. 1978. Occurrence and relative abundance of small mammals associated with riparian and upland habitats along the Columbia River. M.S. thesis. Oregon State University, Corvallis.

Hirsch, L., and J. Riefler. 1981. A floristic study of eight vernal pools [Lower Table Rock Preserve, Jackson Co., OR]. The Nature Conservancy, Oregon Field Office, Portland.

Hitchcock, A.S. [1950] 1971. *Manual of the Grasses of the United States.* 2nd ed. Revised by A. Chase. Reprint (2 vols.). Dover Publications, New York.

Hitchcock, C.L., and A. Cronquist. 1973. *Flora of the Pacific Northwest.* University of Washington Press, Seattle.

Hitchcock, C.L., A. Cronquist, M. Ownbey and J.W. Thompson. 1955–69. *Vascular Plants of the Pacific Northwest.* 5 vols. University of Washington Press, Seattle.

Horn, E.L. 1972. *Wildflowers.* Part 1, *The Cascades.* The Touchstone Press, Beaverton, Oregon.

Horowitz, E.L. 1978. *Our Nation's Wetlands.* U.S. Government Printing Office.

Hotchkiss, N. 1972. *Common Marsh, Underwater and Floating-leaved Plants of the United States and Canada.* Dover Publications, New York.

Jepson, W.L. 1909–1943. *A Flora of California.* 3 vols. Associated Students Store, University of California, Berkeley.

Johnson, D.M., R.R. Petersen, D.R. Lycan, J.W. Sweet and M.E. Neuhaus. 1985. *Atlas of Oregon Lakes.* Oregon State University Press, Corvallis.

Kagan, J. 1983. Willamette Valley wetlands and wet prairies. *Bulletin of the Native Plant Society of Oregon* 16(8): 12.

Kauffman, J.B., and K.P. Connelly. 1988. Ecological effects of fire in Willamette Valley prairies: effects on rare and endangered plants and ecosystems. Progress report 1. The Nature Conservancy, Oregon Field Office, Portland.

Kentula, M., and J.A. Kusler. 1990. *EPA Wetland Creation and Restoration: The Status of the Science.* U.S. Environmental Protection Agency, Corvallis, Oregon.

Kozloff, E.N. 1976. *Plants and Animals of the Pacific Northwest.* University of Washington Press, Seattle.

Kranz, R.D., and J.E. Richter. 1980. A stewardship master plan of Lower Table Rock Preserve. The Nature Conservancy, Oregon Field Office, Portland.

Lawrence, G.H.M. 1951. *Taxonomy of Vascular Plants.* Macmillan Publishing Co., New York.

———. 1955. *An Introduction to Plant Taxonomy.* Macmillan Publishing Co., New York.

Lippert, B.E., and D.L. Jameson. 1964. Plant succession in temporary ponds of the Willamette Valley, Oregon. *American Midland Naturalist* 71: 181–197.

Magee, D. 1981. *Freshwater Wetlands: A Guide to Common Indicator Plants of the Northeast.* University of Massachusetts Press, Amherst.

Marshall, J. 1985. Value assessment of Jackson-Frazier wetland, Benton County, Oregon: a case study. M.S. thesis. Oregon State University, Corvallis.

Mason, H.L. 1957. *A Flora of the Marshes of California.* University of California Press, Berkeley.

Mitsch, W.J., and J.G. Gosselink. 1986. *Wetlands.* Van Nostrand Reinhold, New York.

Moir, W., and P. Mika. 1972. Prairie vegetation of the Willamette Valley, Benton Co., Oregon. Research Work Unit 1251. U.S.D.A. Forest Science Laboratory, Corvallis, Oregon.

Muenscher, W.C. 1944. *Aquatic Plants of the United States.* Handbooks of American Natural History, vol. 4. Ed. A.H. Wright. Comstock Publishing Co., Ithaca, New York.

———. 1975. *Poisonous Plants of the United States.* Collier Books, New York.

Muhlberg, H. 1982. *The Complete Guide to Water Plants: A Reference Book.* EP Publisher, German Democratic Republic.

Munz, P.A. 1964. *Shore Wildflowers of California, Oregon, and Washington.* University of California Press, Berkeley.

Murphy, E.V. 1959. *Indian Uses of Native Plants.* Mendocino County Historical Society, Fort Bragg, California.

Niering, W.A. 1966. *The Life of a Marsh.* McGraw Hill, Toronto.

———. 1991. *Wetlands of North America.* Thomasson-Grant, Charlottesville, Virginia.

Oregon Natural Heritage Data Base. 1987. *Rare, Threatened and Endangered Plants and Animals of Oregon.* Portland.

Otto, N.E., and T.R. Bartley. 1972. *Aquatic Pests on Irrigation Systems: Identification Guide.* Technical Publication, U.S. Department of the Interior, Washington, D.C.

Peck, M.E. 1961. *A Manual of the Higher Plants of Oregon.* 2nd ed. Binford and Mort Publishers, Portland, Oregon.

Pojar, J., and A. MacKinnon. 1994. *Plants of the Pacific Northwest Coast.* Lone Pine Publishing, Edmonton, Alberta.

Prescott, G.W. 1969. *The Aquatic Plants.* Pictured Key Nature Series. Wm. C. Brown Co. Publishers, Dubuque, Iowa.

———. 1980. *How to know the Aquatic Plants.* 2nd ed. Wm. C. Brown Co. Publishers, Dubuque, Iowa.

Raven, P., R.F. Evert and S.E. Eichhorn. 1986. *Biology of Plants.* 2nd ed. Worth Publishers, New York.

Reed, P.B., Jr. 1988. *National List of Plant Species That Occur in Wetlands: 1988 National Summary.* U.S. Fish and Wildlife Service. Biological Report 88 (24).

Reinhold, T.T., and W. Queen. 1974. *Ecology of Halophytes.* Academic Press, New York

Sampson, A.W. [1940] 1972. *Native American Forage Plants.* Reprint of *A Manual of Aquatic Plants.* Fassett, North Carolina.

Sanville, W.D., H.P. Eilers, T.R. Boss and T.G. Pfleeger. 1986. Environmental gradients in northwest freshwater wetlands. *Environmental Management* 10: 125–134.

Savonen, C. 1988. Historical wetlands of the west Eugene study area. Lane Council of Governments, Eugene, Oregon.

Seyer, S.C. 1983. Ecological analysis, Multorpor Fen Preserve, Oregon. The Nature Conservancy, Oregon Field Office, Portland.

Shaw, S.P., and C.G. Fredine. 1956. *Wetlands of the United States, Their Extent and Their Value to Waterfowl and Other Wildlife.* U.S. Fish and Wildlife Service Circular 39.

Siddall, J.L., K.L. Chambers and D.H. Wagner. 1979. *Rare, Threatened and Endangered Vascular Plants in Oregon.* Oregon State Land Board, Salem.

Smith, W.P. 1985. Plant associations within the interior valleys of the Umpqua River Basin, Oregon. *Journal of Range Management* 38: 526–530.

Spellenberg, R. 1979. *The Audubon Society Field Guide to North American Wildflowers.* Alfred A. Knopf, New York.

State of Illinois. 1981. *Aquatic Weeds: Their Identification and Methods of Control.* Fishery Bulletin No. 4. Department of Conservation, Division of Fish and Wildlife Resources.

Steward, A.N., L.J. Dennis and H.M. Gilkey. 1963. *Aquatic Plants of the Pacific Northwest.* Studies in Botany No. 11. Oregon State University Press, Corvallis.

Studola, J. *Encyclopedia of Water Plants.* T.F.H. Publications, Jersey City, New Jersey.

Teal, J., and M. Teal. 1969. *Life and Death of a Salt Marsh.* Ballantine Books, New York.

Thomas, D.W. 1980. Study of the intertidal vegetation of the Columbia River estuary, July–September 1980. Columbia River Estuary Task Force, Astoria, Oregon.

Weldon, L.W., R.D. Blackburn and D.S. Harrison. [1969] 1973. *Common Aquatic Weeds.* U.S.D.A. Agricultural Handbook No. 352. Reprint. Dover Publications, New York.

Weller, M.W. 1981. *Freshwater Marshes.* University of Minnesota Press, Minneapolis.

INDEX TO COMMON AND SCIENTIFIC NAMES

Primary entries are in bold-face type; alternative names and the names of species mentioned only in 'Similar Species' are not.

A

Abies grandis 194
Acer circinatum 194
Agropyron spicatum 146
Agrostis
alba
  var. alba 156
  var. stolonifera 156
capillaris 156
diegoensis 156
stolonifera 156
tenuis 156
alder,
  red 214
  Sitka 214
Alisma
  gramineum 71
    var. angustissimum 71
    var. gramineum 71
  lanceolatum 71
  plantago-aquatica 72
alkali-grass 80
Allium amplectens 100
Alnus
  rubra 214
  sinuata 214
Alopecurus
  aequalis 145
  geniculatus 145
  pratensis 144
Amelanchier alnifolia 177
Anaphalis
  margaritacea 136
Angelica
  arguta 204
  genuflexa 204
angelica,
  kneeling 204
  sharp-tooth 204
Anthoxanthum
  odoratum 147
apple, wild 180
arrowhead, broad-leaf 70
Aruncus
  dioicus 211, 212
  sylvester 211, 212
ash, Oregon 213
aspen, quaking 215
Aster 134
  chilensis 134
    var. hallii 134

curtus 135
hallii 134
subspicatus 135
aster, 134
  common California 134
  Douglas' 134
  Hall's 134
  white-top 135
avens, large-leaf 113
Azolla
  filiculoides 31
  mexicana 31

B

Balsamorhiza
  deltoidea 70, 137
  sagittata 70, 137
balsamroot,
  arrow-leaf 70, 137
  deltoid 70, 137
  Puget 70, 137
barley, meadow 146
barnyard grass,
  large 142
Beckmannia
  syzigachne 153
bedstraw,
  common 130, 199
  small 199
  wall 130
beggarticks,
  devil 139
  leafy 139
  lobed 139
  nodding 78
bentgrass,
  colonial 156
  creeping 156
  leafy 156
betony,
  great 200
  marsh 200
  Mexican 200
  rigid 200
Betula glandulosa 174
Bidens
  cernua 78
  frondosa 139
  tripartita 139
bidens, three-lobed 139
birch, bog 174

bittercress,
  little western 106
  western 40
  Willamette Valley 106
bladderwort,
  common 54
  humped 54
  lesser 54
  mountain 54
blue-eye-grass,
  narrow-leaf 101
blue-joint 151
bluegrass,
  annual 152
  Canadian 206
  fowl 152
  Kentucky 152
  rough 152
bogbean 174
bogcress, yellow 76
Boisduvalia
  densiflora 118
  stricta 118
Brasenia schreberi 36
brittlewort 46, 49
Briza
  maxima 154
  minor 154
Brodiaea
  elegans 100
  hyacinthina 100
brodiaea,
  elegant 100
  hyacinth 100
brome, California 148
Bromus
  briziformis 154
  carinatus 148
brooklime, American 64
brown-leaf, floating 41
bugleweed,
  American 122
  northern 122
bulrush,
  flat-sedge 87
  hard-stem 89
  small-fruited 87
  soft-stem 88
bur-marigold, nodding 78
bur-reed,
  giant 67
  simple-stem 67

buttercup,
  aquatic 55
  celery-leaf 59
  creeping 112
  dwarf 109
  hooked 111
  Lobb's water 55
  spearwort 59
  straight-beak 110
  water-plantain 109
  western 112
  white water 55
  yellow water 55, 59

C

Calamagrostis
  canadensis 151
calico-flower, blue 131
Callitriche 60
  hermaphroditica 61
  heterophylla 61
  stagnalis 61
  verna 61
Calocedrus decurrens 217
Caltha asarifolia 174
camas,
  common 98
  great 98
Camassia
  leichtlinii 98
  quamash 98
canary-grass, reed 143
candy-flower 198
Cardamine
  occidentalis 40
  oligosperma 106
  penduliflora 106
Carex
  amplifolia 87
  aperta 82
  aquatilis var. dives 174
  arthostachya 161
  cusickii 84
  cusickii 158
  densa 158
  deweyana 207
  feta 159
  hendersonii 207
  lanuginosa 208
  leporina 158, 159
  lyngbyei 160

ABOUT THE AUTHOR AND PHOTOGRAPHER

JENNIFER GUARD's interest in botany developed while she was pursuing an English degree at Portland State University. Her first college science class focused on botany; the subject immediately captivated her and led her to two years as an assistant botany instructor, a minor in botany and a career in wetland ecology. During her last two college summers, she worked for the U.S. Forest Service conducting wetland, riparian and alpine plant association studies in Mt. Hood and Gifford Pinchot National forests of Oregon and Washington. While pursuing graduate studies at the University of Oregon, Jennifer was recruited and hired by a private engineering corporation to develop an environmental division. This gave Jennifer the opportunity in 1991 to found Wetland Specialties, Inc., an Oregon-based environmental consulting business that offers botanical, wetland and land planning services to municipal, industrial and private organizations throughout the Pacific Northwest.

Through Wetland Specialties, Jennifer promotes her excitement and loving concern for wetlands and botany, and she has designed and managed many jurisdictional wetland studies, wetland inventories, natural resource and land-use studies, botanical surveys and wetland mitigations throughout western Oregon and Washington. She also serves on various citizen's advisory and steering committees and is a frequent lecturer at universities and schools and for public-interest groups. Jennifer values these opportunities to promote education toward the ecological and psychological necessities of preserving our wild and native landscapes. She is a natural resource specialist and a botanist who understands the need to regulate growth and development with careful natural resource planning.

TRYGVE STEEN is a professor of biology at Portland State University. He has been on the faculty there for the past 25 years, and he also teaches in both the environmental studies and photography programs. He has a master's degree from U.C. Berkeley and a PhD in biology from Yale. As a biologist, photographer and educator, Trygve has a special interest in presenting the extraordinary beauty, biodiversity and distinctive features of the ecosystems of the Pacific Northwest through images that range from aerials to extreme close-ups. His photographs have been extensively published in Europe and the United States by the Sierra Club, the National Audubon Society and the National Geographic Society, among others. They have also been used in museum exhibits throughout the country. Trygve has developed environmental multi-media productions, which have been shown internationally. A major goal of his work is to educate and motivate others to preserve essential features of natural ecosystems and thereby make progress toward living in harmony with the environment that sustains us.

MORE GREAT BOOKS ON THE WORLD OUTDOORS!

Trees, Shrubs & Flowers to Know in Washington & British Columbia

By Chess P. Lyons and Bill Merilees

This all-new, revised edition of a classic guide that has been in print for more than 40 years brings Chess Lyons' lifetime of field experience to a whole new generation. Designed for beginner to intermediate use, this user-friendly guide enables any hiker, naturalist or armchair enthusiast to identify more than 600 common trees, shrubs and wildflowers of this region.

376 pp / 446 color photographs / Softcover / 59 maps / $15.95 / ISBN 1-55105-062-5

Plants of the Pacific Northwest Coast

Edited by Andy MacKinnon and Jim Pojar

This field guide describes 794 species of trees, shrubs, wildflowers, grasses, sedges, rushes, ferns, liverworts, mosses and lichens. More than 1100 color photographs and 797 color maps make this THE most comprehensive botanical field guide for the coastal area from Alaska to Oregon.

528 pp / 1100 color photographs / 900 B & W illustrations / Softcover / 797 color maps / $19.95 / ISBN 1-55105-040-4

Mosses, Lichens and Ferns of Northwest North America

By Dale Vitt, Janet Marsh and Robin Bovey

This photographic field guide covers the area from southern Oregon to Alaska, and from the Pacific Coast to Montana and Saskatchewan. Over 370 species are featured, each illustrated with full-color photographs and distribution maps. Many similar species are mentioned along with habitat and identification keys. This is the perfect book to help you explore the beautiful world in your own backyard.

296 pp / over 400 color photographs / Softcover / $24.95 / ISBN 1-919433-41-3

Mushrooms of Northwest North America

By Helene Schalkwijk-Barendsen

This field guide covers the area from Alaska south to California, from the Pacific Coast to the Dakotas. Information on identification, edibility and potential hazards, habitat and distribution is included, and the author's illustrations make this guide a beautiful and easy-to-use reference for the amateur or professional naturalist.

416 pp / 560 colour illustrations / Softcover / $19.95 / ISBN 1-55105-046-3

Hiking the Ancient Forests of British Columbia and Washington

By Randy Stoltmann

This book celebrates the trails through our few remaining old-growth forests. The hikes described in this book cover the ancient forests of the North Cascades, Mount Rainier and Olympic Peninsula in Washington State, and the Lower Mainland and Vancouver Island in British Columbia.

192 pp / b&w photos & illustrations throughout / Softcover / $15.95 / ISBN 1-55105-045-5

Lone Pine Publishing

1901 Raymond Ave. SW, Suite C
Renton, Washington
USA 98055
Phone (425) 204-5965
Fax (425) 204-6036

202A, 1110 Seymour Street
Vancouver, British Columbia
Canada V6B 3N3
Phone (604) 687-5555
Fax (604) 687-5575

206, 10426 – 81 Avenue
Edmonton, Alberta
Canada T6E 1X5
Phone (403) 433-9333
Fax (403) 433-9646